Rose Lake Lib

COASTAL WETLANDS

Edited by

Harold H. Prince
Department of Fisheries and Wildlife

and

Frank M. D'Itri
Institute of Water Research
and
Department of Fisheries and Wildlife

Michigan State University
East Lansing, Michigan

LEWIS PUBLISHERS, INC.

Library of Congress Cataloging-in-Publication Data

Main entry under title:

Coastal wetlands.

 Proceedings of the First Great Lakes Coastal Wetlands
Colloquium, Nov. 5–7, 1984, East Lansing, Mich., co-
sponsored by the National Sea Grant College Program and
Environment Canada.
 Bibliography: p.
 Includes index.
 1. Wetland ecology—Great Lakes Region—Congresses.
2. Wetlands—Great Lakes Region—Congresses. I. Prince,
Harold H. II. D'Itri, Frank M. III. Great Lakes
Coastal Wetlands Colloquium (1st : 1984 : East Lansing,
Mich.) IV. National Sea Grant College Program (U.S.)
V. Canada. Environment Canada.
QH104.5.G7C6   1985   574.5′26325′0977   85-19914
ISBN 0-87371-052-5

LEWIS PUBLISHERS, INC.
121 South Main Street, P.O. Drawer 519, Chelsea, Michigan 48118

PRINTED IN THE UNITED STATES OF AMERICA

PREFACE

This book represents the proceedings of the first
"Great Lakes Coastal Wetlands Colloquium" (November 5-7,
1984; East Lansing, Michigan). The theme was "Natural
and Manipulated Water Levels in Great Lakes Wetlands."
This material constitutes both Great Lakes wetlands and
the state of understanding about them. It is intended
to provide fisheries and wildlife biologists, ecologists,
aquatic resource managers and planners and environmental
scientists information about the coastal wetlands in regard
to eight priority areas. The colloquium and publication
of the proceedings were cosponsored by Sea Grant Program
and Environment Canada. The material presented by the
authors does not necessarily represent the official views
of these agencies.

Objectives of the colloquium were:

1. To provide a forum for the exchange of current
   information on Great Lakes coastal wetlands,
   relating in particular to water levels.

2. To establish a network of wetland ecologists
   and managers in the Great Lakes region.

3. To publish an integrative set of invited and
   contributed papers on Great Lakes coastal wetlands.

4. To develop a set of research priorities for
   Great Lakes wetlands as a base for future research.

Our judgment is that the colloquium was successful.
We are pleased that this book will endure as testimony
to that assessment and hope that it will stimulate future
syntheses along these lines. There are nine papers that

III

focus around the colloquium theme. They present an overview of the subject, lay out some priorities for research to provide needed information for the future, and have an invited discussion section. The other seven papers are reports of current work that contribute to our knowledge of the processes and management of coastal wetlands.

Harold H. Prince
Michigan State University
East Lansing, Michigan

H.J. Harris
University of Wisconsin-
  Green Bay
Green Bay, Wisconsin

Thomas H. Whillans
Trent University
Peterborough, Ontario

Joachim T. Moenig
Environment Canada
Ottawa, Ontario

Harold Prince   Frank D'Itri

Harold H. Prince is a Professor of Wildlife in the Department of Fisheries and Wildlife at Michigan State University in East Lansing, Michigan. He holds a Ph.D. in Wildlife Ecology from Virginia Polytechnic Institute and State University.

Professor Prince has been directing research on interaction of genetic and environmental parameters on reproduction and survival of birds with a special focus on energetics, habitat selection and biocides. He has authored or co-authored over thirty scientific articles on a variety of topics.

Frank M. D'Itri is a Professor of Water Chemistry in the Institute of Water Research and Department of Fisheries and Wildlife at Michigan State University in East Lansing, Michigan. He holds a Ph.D. in analytical chemistry with special emphasis on the transformation and translocation of phosphorus, nitrogen, heavy metals and hazardous organic chemicals in the environment.

Professor D'Itri is listed in <u>American Men of Science, Physical and Biological Science</u>. He is the author of <u>The Environmental Mercury Problem</u>, co-author of <u>Mercury Contamination</u> and <u>An Assessment of Mercury in the Environment</u>. He is the editor of <u>Wastewater Renovation and Reuse, Land Treatment of Municipal Wastewater: Vegetation Selection and Management, Municipal Wastewater in Agriculture, Acid Precipitation: Effects on Ecological Systems, Artificial Reefs: Marine and Freshwater Applications, A Systems Approach to Conservation Tillage</u> and co-editor of <u>PCBs: Human and Environmental Hazards</u>. In addition, he is the author of more than forty scientific articles on a variety of environmental topics.

Dr. D'Itri has served as chairperson for the National Research Council's panel reviewing the environmental effects of mercury and has conducted numerous symposia on the analytical problems related to environmental pollution. He has been the recipient of fellowships from Socony-Mobil, the National Institutes of Health, the Rockefeller Foundation and the Japan Society for the Promotion of Science.

To the

NATIONAL SEA GRANT COLLEGE PROGRAM

National Oceanic and Atmospheric Administration
U.S. Department of Commerce
Rockville, Maryland

the

MICHIGAN SEA GRANT COLLEGE PROGRAM

Michigan State University
East Lansing, Michigan

The University of Michigan
Ann Arbor, Michigan

and the

ENVIRONMENTAL CONSERVATION SERVICE

Environment Canada
Ottawa, Ontario

ACKNOWLEDGMENTS

The success of this endeavor was due to the interest and dedication of the participating individuals and organizations. Primary among these was the Michigan Sea Grant College Program in cooperation with the National Sea Grant College Program, National Oceanic and Atmospheric Administration (NOAA), U.S. Department of Commerce, and the Environmental Conservation Service, Environment Canada which sponsored and provided the necessary financial support for the conference and publication of this book. Special thanks are extended to Dr. Niles R. Kevern, Associate Director of the Michigan Sea Grant College Program, for his encouragement and support. A grateful acknowledgment also goes to all of the authors whose research, in addition to contributions of time, effort, and counsel made this volume possible.

In the Institute of Water Research at Michigan State University, our special thanks are extended to Dr. Jon F. Bartholic for his support and encouragement. In the preparation of this manuscript, we acknowledge and thank Ms. Dara J. Philipsen for many editorial suggestions that contributed materially to the overall style.

FOREWORD

The Laurentian Great Lakes probably have a place on even the most rudimentary mental map of North America. A waterway into the center of a continent; the largest accumulation of freshwater in the world; a magnificent and valuable ecosystem; notorious pollutants--many are the reasons for the high public profile. It is unlikely that one of the major reasons has been recognition of resources related to the Great Lakes' coastal wetlands. Perhaps the time is due.

Wetlands have long been recognized as locally or regionally important within the Great Lakes Basin. Only recently have they slipped into binational discussions. Just as stewardship of Great Lakes water quality has become a binationally shared responsibility, so might that of coastal wetlands. In fact, these resources can be linked.

Great Lakes water quality can be affected by wetlands in two major ways. Contaminants, especially nutrients, can be uptaken by wetland plants and rendered relatively inaccessible to other organisms. Rooted macrophytes can also serve to anchor sediments, thereby reducing resuspension of sediment--one of the major sources of pollution in many areas. Thus, although wetlands are not addressed explicitly in the 1978 Canada-U.S. Water Quality Agreement, they are clearly germane to the intentions of that Agreement.

Wetlands are "quietly" addressed by the International Great Lakes Fishery Commission in terms of fish habitat or as components of key systems. Some of its policies and funding have been or could be applied to wetland systems. The Migratory Bird Treaty Act does the same for waterbirds by recognizing coastal wetlands as key migrational and breeding habitats. This has been the basis for some governmental study and management in major wetlands around the Great Lakes.

It is fair to say, however, that in spite of general scientific opinion that wetlands are important to Great Lakes ecosystems, they represent one of the least well understood parts of those systems. Moreover, they are greatly diminished in extent and quality along most moderately settled shorelines. Still, in 1981, around the heavily settled lower Great Lakes (Ontario, Erie, and St. Clair) about 61,480 hectares of coastal wetland remained according to the International Lake Erie Regulation Study Board.

These and other wetlands adjacent to the Great Lakes are distinctive in comparison to wetlands around the world. Perhaps their most remarkable characteristic is persistence under the influence of substantial water level fluctuation. Certainly, salt marshes may be subjected to regular tides. It is unusual, however, for freshwater wetlands to undergo such large and irregular daily seiches; long-term hydrologic cycles and storms are typical of the Great Lakes.

As water consumption and water level manipulation continue to increase within the Great Lakes Basin and as thirsty eyes from mid-western North America in particular evaluate the Great Lakes as a solution to an impending water supply crisis, then the need for a detailed and integrated understanding of Great Lakes wetlands and water levels will become more apparent.

Thomas H. Whillans
Trent University
Peterborough, Ontario

COASTAL WETLANDS
CONTRIBUTORS AND PARTICIPANTS

C. Davison Ankney, University of Western Ontario, London, Ontario

Constance Athman, Forest Service Hiawatha National Forest, 2525 North Lincoln Road, Escanaba, Michigan 49829

*John P. Ball, Department of Zoology, University of Guelph, Guelph, Ontario N1G 2W1

Theodore R. Batterson, Assistant Professor, Department of Fisheries and Wildlife, Michigan State University, East Lansing, Michigan 48824

*Donald L. Beaver, Associate Professor, Department of Zoology and The Museum, Michigan State University, East Lansing, Michigan 48824

David W. Bolgrien, Lawrence University, Appleton, Wisconsin 54912

*Eleanor Z. Bottomley, Wetlands Biologist, Canadian Wildlife Service, 1725 Woodward Drive, Ottawa, Ontario K1A 0E7

Daniel C. Brazo, Lake Michigan Research Specialist, Indiana Department of Natural Resources, Michigan City, Indiana 46360

James M. Breitenbach, Department of Fisheries and Wildlife, Michigan State University, East Lansing, Michigan 48824

Brenda Brobst, Halton Region Conservation Authority, 310 Main Street, Milton, Ontario

*Thomas M. Burton, Professor, Departments of Zoology and Fisheries and Wildlife, Michigan State University, East Lansing, Michigan 48824

x

John Cantlon, Vice President, Research and Graduate Studies, Michigan State University, East Lansing, Michigan 48824

*Saralee Chubb, Department of Fisheries and Wildlife, Michigan State University, East Lansing, Michigan 48824

Daryl Cowell, Environmental Canada, 25 St. Clair Avenue, E., Toronto, Ontario M4T 1M2

John Craig, Department of Fisheries and Wildlife, Michigan State University, East Lansing, Michigan 48824

Janice Doane, M.T.R.C.A., 5 Shoreham Drive, North York, Ontario M3N 1S4

Ronald D. Drobney, Center for Environmental and Estuarine Studies, University of Maryland, Frostburg, Maryland 21532

*Gary Fewless, University of Wisconsin-Green Bay, Green Bay, Wisconsin 54302

Millie Flory, Michigan Sea Grant College Program, 4109 Institute of Science & Technology Building, University of Michigan, Ann Arbor, Michigan 48109

*Theodore R. Gadawski, Ducks Unlimited Canada, 240 Bayview Drive, Unit 10, Barrie, Ontario

*James W. Geis, College of Environmental Science and Forestry, State University of New York, Syracuse, New York 13210

*Valanne Glooschenko, Wetland Habitat Coordinator, Wildlife Branch, Ontario Ministry of Natural Resources, Queen's Park, Toronto, Ontario M7A 1W3

James R. Goodheart, Department of Natural Resources, Wildlife Branch, Queens Park, Toronto, Ontario

William S. Graham, National Sea Grant Office, Washington, D.C.

Laura Grantham, Department of Fisheries and Wildlife, Michigan State University, East Lansing, Michigan 48824

Richard H. Greenwood, U.S. Fish and Wildlife Service, 1405 South Harrison Road, Room 301, East Lansing, Michigan 48823

*Glenn Guntenspergen, Research Associate, Department of Biological Sciences, P.O. Box 413, University of Wisconsin-Milwaukee, Milwaukee, Wisconsin 53201

*David E. Hammer, Wetlands Ecosystem Research Group, Department of Chemical Engineering, University of Michigan, Ann Arbor, Michigan 48109

*H.J. Harris, Sea Grant Program, University of Wisconsin-Green Bay, Green Bay, Wisconsin 54302

Mark H. Hart, Department of Fisheries and Wildlife, Michigan State University, East Lansing, Michigan 48824

Nancy Holm, Illinois State Geological Survey, 615 East Peabody Drive, Champaign, Illinois 61820

J. A. Holmes, University of Toronto, 21 Hollybrook Crescent, Willowdale, Ontario M2J 2H5

Robert Humphries, Michigan Department of Natural Resources, 36900 Mouillee Road, Rockwood, Michigan 48173

Martin Jannereth, Michigan Department of Natural Resources, 7726 Cloverhill Drive, Lansing, Michigan 48917

Carlton R. Jarvis, Michigan Department of Natural Resources, 672 Westchester Drive, Caro, Michigan 48723

Eugene Jaworski, Eastern Michigan University, Ypsilanti, Michigan 48197

Carey Johnson, Michigan Department of Natural Resources, P.O. Box 30028, Lansing, Michigan 48909

David L. Johnson, Ohio State University, Columbus, Ohio 43210

Joe Johnson, Kellogg Biological Station, Hickory Corners, Michigan 49060

John H. Judd, Michigan Sea Grant, 334 Natural Resources Building, Michigan State University, East Lansing, Michigan 48824

*Robert H. Kadlec, Wetlands Ecosystem Research Group, Department of Chemical Engineering, University of Michigan, Ann Arbor, Michigan 48109

Ernest N. Kafcas, Wildlife Habitat Biologist, 33135 South River Road, Mt. Clemens, Michigan 48045

*Jeffrey M. Kaiser, Consulting Biologist, Mississauga, Ontario

Ted Kaiser, National Oceanic and Atmospheric Administration, 2300 Washtenaw Avenue, Ann Arbor, Michigan 48104

*Richard M. Kaminski, Department of Wildlife and Fisheries, Mississippi State University, P.O. Drawer LW, Mississippi State, Mississippi 39762

*Paul A. Keddy, Department of Biology, University of Ottawa, Ottawa, Ontario K1N 6N5

James C. Kelley, Graduate Student, Department of Fisheries and Wildlife, Michigan State University, East Lansing, Michigan 48824

*Darrell L. King, Professor, Department of Fisheries and Wildlife, Michigan State University, East Lansing, Michigan 48824

George Knoecklein, 28 Club House Drive, Kincheloe, Michigan 49788

Margaret T. Kolar, U.S. Fish and Wildlife Service, 1405 South Harrison Road, East Lansing, Michigan 48823

Keith D. Kraai, U.S. Fish and Wildlife Service, 1405 South Harrison Road, East Lansing, Michigan 48823

*Charles R. Liston, Associate Professor, Department of Fisheries and Wildlife, Michigan State University, East Lansing, Michigan 48824

Barry Loper, 3101 Manley Drive, Lansing, Michigan 48910

John G. Lyon, Department of Civil Engineering and the Ohio Sea Grant Program, Ohio State University, Columbus, Ohio 43210

Lynwood A. MacLean, U.S. Fish and Wildlife Service, 3990 East Broad Street, Columbus, Ohio 43215

D. K. Martin, University of Toronto, 21 Hollybrook Crescent, Toronto, Ontario M2J 2H5

*Gary B. McCullough, Canadian Wildlife Service, 152 Newbold Court, London, Ontario N6E 1Z7, Canada

Douglas McLaughlin, University of Wisconsin, 2247 East Shore Drive, Green Bay, Wisconsin 54302

Clarence D. McNabb, Professor, Department of Fisheries and Wildlife, Michigan State University, East Lansing, Michigan 48824

*Martin K. McNicholl, Executive Director, Long Point Bird Observatory, P.O. Box 160, Port Rowan, Ontario N0E 1M0

James W. Merna, Michigan Department of Natural Resources, Museums Annex, Ann Arbor, Michigan 48109

Joachim T. Moenig, Water Planning and Management Branch, Inland Waters Directorate, Environment Canada, 1177 Beaujolais, Ottawa, Ontario K1A 0E7

*Henry R. Murkin, Delta Waterfowl and Wetland Research Station, RR1, Portage la Prairie, Manitoba R1N 3A1

Robert Murphy, Environmental Studies, University of Wisconsin, Green Bay, Wisconsin 54301

Thomas L. Nederveld, Michigan Department of Natural Resources, 906 Miami Drive, Mason, Michigan 48854

*Jeffrey W. Nelson, Ducks Unlimited Canada, Winnipeg, Manitoba R3T 2E2

Mike Oldham, Essex Region Conservation Authority, 360 Fairview Avenue, W., Essex, Ontario N8M 1Y6

Bob Owens, U.S. Fish and Wildlife Service, 1405 South Harrison Road, East Lansing, Michigan 48823

Robert D. Pacific, U.S. Fish and Wildlife Service, 1405 South Harrison Road, East Lansing, Michigan 48823

Paul I. Padding, Department of Fisheries and Wildlife, Michigan State University, East Lansing, Michigan 48824

*Nancy J. Patterson, Federation of Ontario Naturalists, 355 Lesmill Road, Don Mills, Ontario M3B 2W8

Timothy C. Payne, Michigan Department of Natural Resources, 2455 North Williams Lake Road, Pontiac, Michigan 48054

*Harold H. Prince, Professor, Department of Fisheries and Wildlife, Michigan State University, East Lansing, Michigan 48824

*Henry A. Regier, Institute for Environmental Studies, University of Toronto, Toronto, Ontario M5S 1A4

John Reynolds, Consumers Power Company, 212 West Michigan, Jackson, Michigan 49201

*Anton A. Reznicek, University of Michigan Herbarium, Ann Arbor, Michigan 48109

*Sumner Richman, Lawrence University, Appleton, Wisconsin 54912

Ray Rusteum, Michigan United Conservation Clubs, 2101 Wood Street, Lansing, Michigan 48912

Steven G. Sadewasser, Michigan Department of Natural Resources, P.O. Box 30028, Lansing, Michigan 48909

*Paul E. Sager, ES-317, University of Wisconsin-Green Bay, Green Bay, Wisconsin 54301-7001

Wayne Schmidt, Michigan United Conservation Clubs, 2101 Wood Street, Lansing, Michigan 48912

Carol Siegley, Ohio State University, 1528 Kenny Road, Columbus, Ohio 43212

*Christopher E. Smith, Ducks Unlimited Canada, Box 1799, The Pas, Manitoba R9A 1L5

David W. Smith, Hunter and Associates, Toronto, Ontario

William S. Snyder, Ohio State University, Columbus, Ohio 43210

Terry Sowden, Ecologistics LTD, 50 Westmount Road, Suite 225, Waterloo, Ontario N2L 2R5

R. J. Steedman, University of Toronto, 21 Hollybrook Crescent, Willowdale, Ontario M2J 2H5

Becky Steinbach, Michigan Department of Natural Resources, 225 East Spruce, St. Charles, Michigan 48655

T. Stevenson, University of Toronto, 21 Hollybrook Cresecent, Willowdale, Ontario M2J 2H5

*William W. Taylor, Associate Professor, Department of Fisheries and Wildlife, Michigan State University, East Lansing, Michigan 48824

Donald L. Tilton, Smith, Hinchman and Grylls Associates, Inc., Detroit, Michigan

Pamela Tyning, Department of Fisheries and Wildlife, Michigan State University, East Lansing, Michigan 48824

Martha Walter, Michigan Sea Grant College Program, 4117A Institute of Science & Technology Building, University of Michigan, Ann Arbor, Michigan 48109

*Patricia B. Weber, Eastern Michigan University, Ypsilanti, Michigan 48197

*Thomas H. Whillans, Environmental and Resource Studies, Trent University, Peterborough, Ontario K9J 7B8

John C. White, Lawrence University, Appleton, Wisconsin 54912

Douglas Wilcox, National Park Service, 1100 North Mineral Springs Road, Porter, Indiana 46304

Donald C. Williams, Corp. of Engineers, P.O. Box 1027, Detroit, Michigan 48231

Charles L. Wolverton, Michigan Department of Natural Resources, 12290 170th Avenue, Big Rapids, Michigan 49307

Jim Wood, Michigan United Conservation Clubs, 2101 Wood Street, Lansing, Michigan 48912

John Wuycheck, Michigan Department of Natural Resources, P.O. Box 30028, Lansing, Michigan 48909

*Authors of papers published in this book

COASTAL WETLANDS
CONTENTS

CHAPTER 1

THE EFFECTS OF WATER LEVEL FLUCTUATIONS
ON GREAT LAKES COASTAL MARSHES

Thomas M. Burton
Departments of Zoology and Fisheries and Wildlife
Michigan State University
East Lansing, Michigan 48824

STRUCTURE AND FUNCTION OF GREAT LAKES COASTAL MARSHES

Many of the wetlands within the Great Lakes Basin have already been converted to other uses. For example, 47 percent or 7.5 of 16 million hectares of wetlands had been destroyed in Michigan, Minnesota and Wisconsin by 1980 (Tiner, 1984). Tiner (1984) cited estimates by others that indicated that an additional 8,100 hectares of wetlands were being lost each year from these three states. Even so, these three states account for 77 percent of the total wetland areas in glaciated regions of the United States (Jaworski and Raphael, 1979). Most of these wetlands are inland with only a small percentage classified as coastal wetlands. For example, 3.3 percent of Michigan's 1.3 million hectares or 42,840 hectares were classified as coastal wetlands by Jaworski et al. (1979).

These wetlands are often considered to be modulators of events between land and water. Some of the functions ascribed to them include: (1) acting as a natural filter to protect the water quality of the Great Lakes from nutrients and toxic materials; (2) acting as flood storage areas to reduce the magnitude of flood damage; (3) acting as areas of concentrated primary and secondary production which may serve as food chain support for near-shore Great Lakes communities; (4) acting as recharge areas for groundwater; and (5) serving as habitat and/or nursery areas for fish, mammals, game and non-game birds as well as invertebrates and ectothermic vertebrates.

The fish, wildlife and recreational value of Great Lakes coastal wetlands are about $1,800/ha (based on Jaworski et al.'s, 1979, estimate adjusted for inflation at a rate of 6 percent per year). It is very difficult to assign a dollar value to the other functions listed above, but the value of marshes should be considerably in excess of Jaworski et al.'s estimate.

All of the functions listed are based on general concepts and are supported by few quantitative data, especially data specific for Great Lakes marshes. In fact, one cannot even be sure whether individual wetlands perform any or all of these functions or not. Sather and Smith (1984) provided an excellent summary of some of these functions and values for marshes in general, while Tilton and Schwegler (1979) provided an overview for wetland habitats in the Great Lakes Basin. Both summaries emphasized the lack of quantitative data and the often contradictory data in the literature. If we expect to examine the impacts of water level fluctuations on Great Lakes coastal marshes, the lack of data becomes even more acute.

We must be cautious not to over-emphasize published generalizations based on only a few studies. For many years, for example, coastal wetlands were touted as the source of organic carbon providing the basis of the food chain for the rich and diverse near-shore fishery of the Atlantic Coast. Nixon (1980) reviewed the evidence for this widely accepted belief and found very little evidence of its veracity. The idea originated with the "coastal outwelling" concept espoused by E. P. Odum in 1968 based on the limited salt marsh energy flux data of Teal published in 1962 (see Nixon, 1980, for review of the origins of the concept).

According to this concept, Spartina and other plants in coastal marshes produce large amounts of organic carbon which are exported or "outwelled" to estuarine and near-shore areas. Sediment-derived nutrients, especially phosphorus, are also taken up by Spartina and outwelled to adjacent areas. This outwelled carbon and phosphorus form much of the energy supporting higher trophic levels in these near-shore areas in a manner analogous to highly productive "upwelling" areas where high oceanic productivity is driven by enriched bottom water surfacing due to oceanic circulation patterns.

Nixon (1980) conducted an extensive review of the salt marsh literature and concluded the following:

"There may be an export of organic carbon from many tidal marshes, and the export may provide a carbon supplement equivalent to a significant fraction of the open water primary production in many areas of the South. It is even possible that this carbon may contribute measurably to the standing crop of organic carbon in the water at any one time. But it does not appear to result in any greater production of finfish or shellfish than is found in other coastal areas without salt marsh organic supplements."

Nixon (1980) also concluded that marshes were not major sources or sinks for coastal marine nutrient cycles and demonstrated that our present knowledge of the role of salt marshes in trace metal budgets, nutrient cycles or organic matter budgets for estuaries suggests a very limited role for marshes for these systems. These conclusions, based on an extensive review of the literature, are opposite most generalizations found in the literature.

Despite the paucity of data supporting the "outwelling" hypothesis, we see the same general concept suggested quite often as a rationale for preserving freshwater wetlands. Even if Nixon's (1980) conclusions are correct for salt marshes, they may be incorrect for freshwater systems such as those of the Great Lakes. Much of the original impetus for re-examining the outwelling hypothesis came from E. B. Haines' carbon-13/carbon-12 ratio data for food sources of estuarine organisms (see Nixon, 1980, for review of several applicable papers by Haines). Her data suggested that estuarine organisms do not use Spartina fragments directly for food. Her studies of Georgia estuaries were corroborated recently for Alaskan estuarine areas (Schell, 1983). However, Schell found that freshwater organisms of tundra ponds and lakes were heavily dependent on peat accumulated from terrestrial areas. He suggested that lack of use of this peat in the near-shore areas of the Beaufort Sea was due to the lack of functionally analogous abundant marine detritivores. The dominant invertebrates of the much better studied Atlantic freshwater tidal marshes are detritivores including oligochaete worms, amphipods, snails and insect larvae (Simpson et al., 1983). Most of the vegetation produced in these wetlands enters the food chain via detrital pathways. The near-shore areas of the Great Lakes support similar freshwater detritivores. Thus, the outwelling hypothesis may be applicable to Great Lakes coastal marshes even through it does not seem to apply to salt marsh systems. The point is that few, if any, quantitative data exist on this topic.

Data to support the other suggested functions are just as tenuous. The nutrient removal function of marshes may occur for certain wetlands while others simply act as flow-through systems for many nutrients (see reviews by Whigham and Bayley, 1979; Burton, 1981; Sather and Smith, 1984). Likewise, the suggested effect on flood storage is based primarily on an analysis by Novitzki (1979) of flood peaks in watersheds in Wisconsin with significant wetlands and small lakes in the basin compared to basins with few wetlands and upland lakes. He showed that reduction in flood flow was related to percentage of a catchment in wetlands or lakes. Any individual wetland may or may not perform this function, and almost no data exist for individual wetlands (Carter et al., 1979). Other suggested hydrologic values such as groundwater recharge or increased stream base flow probably seldom, if ever, occur; and wetlands may actually decrease these parameters (Carter et al., 1979).

The Great Lakes coastal marshes have received very little study compared to Atlantic coast salt marshes. Nixon (1980) documented a wide variability in results of salt marsh studies. We may expect the same variability in Great Lakes coastal marsh studies. Even though any generalizations at this time may be premature, it is worthwhile to take stock of our current state of knowledge and suggest future research needs. The lack of quantitative data documented above suggest the need for a wide variety of quantitative studies. The papers in this volume seek to reveiw current knowledge or report findings of specific research projects. The underlying theme is the role of water level fluctuation on the structure and function of Great Lakes coastal marshes. I will review some of the consequences of water level fluctuations on biogeochemical cycles within the coastal marshes in the remainder of this paper.

WATER LEVEL EFFECTS ON GREAT LAKES MARSHES

The present 7-10 year cycle of water level fluctuation results in low periods in lake level which are about 1.75 m lower than the high (Kelley et al., 1985). The difference between low and high water can have profound effects on the plant communities of coastal marshes (Harris et al., 1981; Kelley et al., 1985). At high water level, much of the marsh area becomes open water with as much as 50 percent of the emergent zones (Typha, Sparghanium) and sedge meadows (Carex, Calamogrostis) being converted to open water (Jaworski et al., 1979). At low water levels, open water decreases from almost 50 percent of wetland area to about 15 percent. These changes are illustrated for Pentwater Marsh on the west coast of Michigan (Oceana County) in Figure 1 (drawn from unpublished data summarized in Kelley et al., 1985). At the low Lake Michigan water level of 175.4 m in 1965, open water was reduced primarily to the main channel and the emergent zone almost ceased to exist (there may have been limited areas of emergent vegetation along the channel, but they could not be differentiated from the available black and white aerial photographs). At high water levels near 177 m in 1975, much of the area of the marsh was occupied by open water/submergent vegetation or emergent vegetation (Figure 1). Even more dramatic changes were documented for the more open (not separated from open water by a "buffer" lake and channel) Green Bay marshes by Harris et al. (1981). These changes in water level and marsh plant community composition are likely to be accompanied by major changes in the biota and biogeochemical cycles. Only limited quantitative data are available on these changes (Kelley et al., 1985). I will use the remainder of this paper to discuss some potential effects of water level fluctuations on the biota and on biogeochemical cycles.

As is apparent from Figure 1, water level fluctuations can dramatically change the plant communities of coastal marshes. These changes may be accompanied by significant changes in biogeochemical cycling. During low water level, more and more of the marsh sediment becomes exposed to aerobic conditions. Decomposition takes place at a more rapid pace under aerobic conditions (Odum, 1979). It is certainly possible that organic matter, nutrients and toxic materials accumulate in a peat or organic substrate under inundated, high productivity conditions and are "flushed" from the system as water level changes. Oxidation of exposed organic matter in sediments during low water level periods would probably result in "flushing" of nutrients such as carbon and nitrogen from the system as runoff over these areas occurs or during initial stages of the next water level increase as re-inundation occurs. Conversely, sediment phosphorus release is related to the depth of the aerobic surface zone (Wetzel, 1975). As sediments are exposed, the aerobic zone at sediment surface should increase substantially. This increased zone should prevent phosphorus from migrating upward as phosphorus is adsorbed on and complexed with ferric oxides and hydroxide (Wetzel, 1975). The oxidized layer also efficiently traps iron and manganese.

As water level increases, inundated areas will support considerable emergent and/or submergent productivity including the associated epiphytic plant productivity. As this material rapidly decomposes, the overlying water dissolved

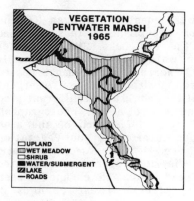
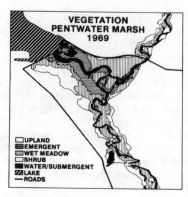
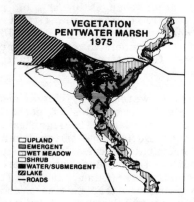
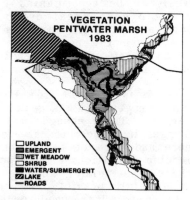

Figure 1. Impact of water level fluctuation on plant communities of the Pentwater Marsh, Michigan, from low (1965) to high water level (1975) with intermediate low (1969) and high levels (1983). Drawn from unpublished data summarized in Kelley et al. (1985).

oxygen concentrations will decrease, especially in winter when oxygen production by plant photosynthesis is limited. Thus, the aerobic surface zone for inundated sediments will decrease or disappear. Under such conditions, accumulated phosphorus, iron and manganese will be reduced to more soluble forms and released to the overlying water. Thus, alternate fluctuations in water level in marshes could result in a situation analogous to that resulting from seasonal re-oxygenation of bottom waters in dimictic eutrophic lakes (i.e., release of phosphorus, iron, manganese and trace metals from the sediments under reduced, high water conditions and accumulation of these materials during oxidizing periods of low water sediment exposure).

It is difficult to predict from knowledge of sediment chemistry the direction of nitrogen retention or release from a marsh. As oxidation of organic matter increases during sediment exposure, increased nitrification should lead to flushing of nitrate from the system. Also, increased denitrification under high water, more anaerobic sediment conditions should result in greater losses of nitrogen gas and more apparent retention or removal of nitrogen by the marsh. In general, one might expect decreased nitrogen retention during low water as organic matter is nitrified and lost from the system with exported water and greater apparent retention under high water conditions as more nitrogen gas is lost due to increased denitrification. However, decomposition rates are less in seasonally inundated wet meadow than in deeper water emergent communities (Kelley et al., 1985). Thus, accumulation in slowly decomposing litter could offset a greater tendency towards nitrification as the aerobic surface zone expands during sediment exposure.

The Great Lakes coastal marshes are unlike other midwestern inland freshwater marshes because of the unique fluctuations in water level due to seiche activity in the lakes. These seiches can result in periodicity in fluctuations which may vary from less than an hour to several hours and may range from a few centimeters to more than a meter. Such tide-like activity profoundly influences import and export of materials and makes mass balance studies difficult to conduct. Frequency and duration of inundation determine substrate characteristics which in turn determines species diversity, primary production, rate of decomposition and uptake and release of nutrients (Gosselink and Turner, 1978; Simpson et al., 1983). Marshes on the west coast of Michigan are usually separated from the main lake by a channel through the dune area and a small lake (Jaworski et al., 1979). The marshes often form where the rivers join the small lake. This type of riverine coastal marsh is illustrated by Pentwater Marsh (Figure 1). Pentwater Marsh still is influenced by seiche activity with periodic fluctuations of 30 mintues to two hours. Flow reversal sometimes occurs during low discharge periods. Even so, this marsh is buffered from seiche activity compared to open marshes such as those of Green Bay (Harris et al., 1981). In these marshes, seiche activity is much more important. Nutrient cycles are difficult to quantify in such "tidal" areas and are likely to vary depending on season, degree of inundation, etc. Simpson et al. (1983) summarized data for Atlantic freshwater tidal marshes and documented a wide variety of responses for nutrients ranging from export from some areas of these marshes to import and storage in other

areas. Wide variability also existed between the three
marshes summarized in their paper. Responses were also
widely variable for various chemical constituents. I
am certain that Great Lakes coastal marshes will vary
as greatly as did the Atlantic freshwater tidal marshes.

These uncertainties about nutrient cycles as water
levels fluctuate emphasize the need for long-term mass
balance budgets for marshes at several sites and the need
for much more research on sediment chemistry under aerobic
and anaerobic conditions as this chemistry is affected
by the associated plant community. Such interactive research
is essentially non-existent. Obviously, what is needed
is much more long-term data on the influence of water
level fluctuation on specific nutrient, trace metal and
organic toxin budgets within marshes. An alternative
approach would be to document in detail specific data
on biogeochemical rates and pathways for a variety of
plant community types and then use these detailed community-
based biogeochemical data to predict changes as water
level and community composition changes.

My colleagues and I (Kelley et al., 1985) attempted
to use the latter approach for Pentwater Marsh for nitrogen
and phosphorus just for uptake and release of these nutrients
by marsh vegetation. Storage of nitrogen and phosphorus
in plant tissue was at a minimum during highest water
in 1975, was highest at intermediate water levels in 1983
and intermediate at lowest water levels in 1965 (see Figure
1 for associated changes in the plant community). Litter
accumulation was greatest under lowest water conditions
due to known slower decomposition rates in sedge meadows
(J. Kelley, unpublished data). Thus, rate of nutrient
release from vegetation was likely lowest at the time
of lowest water level in 1965 (Figure 1) and might suggest
decreased release from the marsh. Of course, vegetation
storage was only one facet of the budget and oxidation
of exposed organic sediments probably more than offset
slower nutrient release from decomposing plant tissue.
Without complete long-term mass balance budgets one cannot
be sure whether the marsh acted as a source or sink for
nutrients during these periods. Preliminary data analyses
for J. Kelley's dissertation suggests that Pentwater Marsh
acted as a slight sink (less than 25 percent retention)
for nitrogen and phosphorus for the 1981-84 period during
intermediate to high water levels. Data from other coastal
riverine marshes suggest that some retention occurs but
that marshes may, at times, act as simple flow-through
systems (Burton, 1981).

The impact of water level changes on some bird and mammal populations has been well documented for inland emergent marshes (Weller, 1979). Few such data are available for the Great Lakes, and almost no data are available for fish populations. Some of the other papers in this volume will summarize on-going research and current knowledge of the effects of water level fluctuation on Great Lakes fish and wildlife. These relatively limited sets of data need to be expanded with more diverse and longer term studies. An area of almost complete lack of data is knowledge of invertebrate and other prey populations in Great Lakes coastal marshes.

In conclusion, it is obvious that most of our ideas about the structure and function of Great Lakes coastal marshes are based on limited data sets and generalities. There is a pressing need for long-term studies at the integrated ecosystem level for a variety of types and locales for Great Lakes marshes. Hopefully, the papers in this volume will provide the overview and identify specific research needs so that more research is initiated on these systems before they are lost or further damaged by pollution.

ACKNOWLEDGMENTS

Parts of this research were supported by Grant R/CW-5 by the Michigan Sea Grant Program, National Oceanic and Atmospheric Administration. I thank James C. Kelley for use of some of his unpublished background data on water level effects on plant communities.

LITERATURE CITED

Burton, T.M. 1981. The effects of riverine marshes on water quality. In: B. Richardson (ed.), Selected Proceedings of the Midwest Conference on Wetland Values and Management. The Freshwater Society, St. Paul, MN, pp. 139-151.

Carter, V., M.S. Bedinger, R.P. Novitzki, and W.O. Wilen. 1979. Water resources and wetlands. In: P.E. Greeson, J.R. Clark, and J.E. Clark (eds.), Wetland Functions and Values: The State of Our Understanding. American Water Resources Association, Minneapolis, MN, pp. 344-376.

Gosselink, J.G. and R.E. Turner. 1978. The role of hydrology in freshwater wetland ecosystems. In: R.E. Good, D.F. Whigham, and R.L. Simpson (eds.), Freshwater Wetlands: Ecological Processes and Management Potential. Academic Press, New York, NY, pp. 63-78.

Harris, H.J., G. Fewless, M. Milligan, and W. Johnson. 1981. Recovery processes and habitat quality in a freshwater coastal marsh following a natural disturbance. In: B. Richardson (ed.), Selected Proceedings of the Midwest Conference on Wetland Values and Management. The Freshwater Society, St. Paul, MN.

Jaworski, E. and C.N. Raphael. 1979. Historical changes in natural diversity of freshwater wetlands, glaciated region of Northern United States. In: P.E. Greeson, J.R. Clark, and J.E. Clark (eds.), Wetland Functions and Values: The State of Our Understanding. American Water Resources Association, Minneapolis, MN, pp. 545-557.

Jaworski, E., C.N. Raphael, P.J. Mansfield, and B. Williamson. 1979. Impact of Great Lakes Water Level Fluctuations on Coastal Wetlands. Final Research Report, U.S. Office of Water Research and Technology, Institute of Water Research, Michigan State University, East Lansing, MI, 351 pp.

Kelley, J.C., T.M. Burton, and W.R. Enslin. 1985. The effects of natural water fluctuations on N and P cycling in a Great Lakes marsh. Wetlands (in press).

Nixon, S.W. 1980. Between coastal marshes and coastal waters - A review of twenty years of speculation and research on the role of salt marshes in estuarine productivity and water chemistry. In: P. Hamilton and K.B. MacDonald (eds.), Estuarine and Wetland Processes with Emphasis on Modeling. Plenum Publishing Corp., New York, NY, pp. 437-525.

Novitzki, R.P. 1979. Hydrologic characteristics of Wisconsin's wetlands and their influence on floods, streamflow, and sediment. In: P.E. Greeson, J.R. Clark, and J.E. Clark (eds.), Wetland Functions and Values: The State of Our Understanding. American Water Resources Association, Minneapolis, MN, pp. 377-388.

Odum, E.P. 1979. The value of wetlands: A hierarchical approach. In: P.E. Greeson, J.R. Clark, and J.E. Clark (eds.), Wetland Functions and Values: The State of Our Understanding. American Water Resources Association, Minneapolis, MN, pp. 16-25.

Sather, J.H. and R.D. Smith. 1984. An Overview of Major Wetland Functions and Values. FWS/OBS-84/18, U.S. Fish and Wildlife Service, Washington, DC, 68 pp.

Schell, D.M. 1983. Carbon-13 and carbon-14 abundances in Alaskan aquatic organisms: Delayed production from peat in aquatic food webs. Science. 219(4588):1068-1071.

Simpson, R.L., R.E. Good, M.A. Leck, and D.F. Whigham. 1983. The ecology of freshwater tidal wetlands. BioScience 33(4):255-259.

Tilton, D.L. and B.R. Schwegler. 1979. The values of wetland habitat in the Great Lakes Basin. In: P.E. Greeson, J.R. Clark, and J.E. Clark (eds.), Wetland Functions and Values: The State of Our Understanding. American Water Resources Association, Minneapolis, MN, pp. 267-277.

Tiner, R.W., Jr. 1984. Wetlands of the United States: Current status and recent trends. National Wetlands Inventory, U.S. Fish and Wildlife Service, Washington, DC, 59 pp.

Weller, M.W. 1979. Wetland habitats. In: P.E. Greeson, J.R. Clark, and J.E. Clark (eds.), Wetland Functions and Values: The State of Our Understanding. American Water Resources Association, Minneapolis, MN, pp. 210-234.

Wetzel, R.G. 1975. Limnology. W.B. Saunders Company, Philadelphia, PA, 743 pp.

Whigham, D.R. and S.E. Bayley. 1979. Nutrient dynamics in freshwater wetlands. In: P.E. Greeson, J.R. Clark, and J.E. Clark (eds.), Wetland Functions and Values: The State of Our Understanding. American Water Resources Association, Minneapolis, MN, pp. 468-478.

CHAPTER 2

ENVIRONMENTAL INFLUENCES ON THE DISTRIBUTION
AND COMPOSITION OF WETLANDS
IN THE GREAT LAKES BASIN

James W. Geis
SUNY College of Environmental Science and Forestry
Syracuse, New York 13210

INTRODUCTION

Wetlands are land-water systems which characterize
shoreline interfaces of most water bodies. Cowardin et
al. (1979) have employed the concept of a continuum of
physical environmental conditions at these aquatic/terrestrial
interfaces to illustrate their definition of wetlands.
To them, wetlands are lands where "the water table is
at, near, or above the land surface long enough to promote
the formation of hydric soils or support the growth of
hydrophytes." The deep water end of the continuum is
marked by the growth limit of emergent macrophytes. It
grades into "deep-water habitats," which are dominated
by submerged aquatic macrophytes. The upland limit is
exceeded when soils are no longer "hydric" in classification,
and the predominating vegetation is terrestrial rather
than hydrophytic.

Our studies along the eastern shoreline of Lake
Ontario and the St. Lawrence River (Gilman, 1976; Geis
and Kee, 1977; Geis, 1979; Ruta, 1981; Geis, Ruta and
Hyduke, 1985) have emphasized the continuity of physical
environmental conditions and the intergradation of dominant
plant species between adjacent wetland and shallow-water
littoral systems. Consequently, we consider a "wetlands
continuum" dominated by aquatic macrophytes, both submerged
and emergent, to represent an ecologically useful concept.
This continuum spans a range of environments from the
deep water limit of submerged aquatic macrophytes to the

.upland contact. The practical delineation of "wetlands" and "deep-water habitats" according to the occurrence of emergent hydrophytes is not seen to be at variance with this concept.

Plant communities within the wetlands continuum are dominated by hydrophytes which are uniquely adapted to growing in hydric soils or sediments and which provide high value habitat for semi-aquatic and aquatic animals. Wetland systems are broad and extensive where this continuum is long, and the degree of environmental change along its extent is gradual. They are limited in occurrence and extent where environmental gradients are steep or truncated.

Coastal wetlands are present throughout the Great Lakes-St. Lawrence River Basin. Their physical settings have developed from landscapes with a common geological history, even though substantial differences may exist in surface deposits and parent materials. They exhibit similarities of physiogonomy, species composition, and community structure as a result of the dominating influence of environmental factors correlated with the water regimes of large bodies of well-mixed fresh water.

FACTORS CONTROLLING WETLAND DISTRIBUTION AND ESTABLISHMENT

A wide variety of physical environmental variables have been demonstrated to influence the distribution and composition of freshwater wetland systems (e.g., Sculthorpe, 1967; van der Valk and Bliss, 1971; Auclair et al., 1973; Millar, 1973). In the Great Lakes Basin, primary importance is attributable to variables correlated with shoreline morphology and the hydrologic regime. These features define the environmental gradients along which the wetlands continuum can develop.

Shoreline Morphology and Physical Protection

Jaworski et al. (1979) state that, "the occurrence, distribution and diversity of coastal wetlands is, in part, determined by the morphology of the coast. Perhaps in no other geographical environment is the relationship between land forms and vegetation so evident." They utilized this perspective to design conceptual models for wetlands in the upper four Great Lakes. A similar conclusion was reached by Geis and Kee (1977) following field surveys along eastern Lake Ontario and the St. Lawrence River in New York State.

Four broad categories of wetland systems are described below on the basis of shoreline morphology, the degree of physical protection from wind and wave action provided by shoreline structures, and the influence of lake water levels on wetland hydrology.

Barrier and Lagoon Systems. The majority of Great Lakes wetlands develop in shallow depressional areas called lagoons or flood ponds which occur immediately landward of the shoreline edge. Glacial drift materials of varying texture usually form the upland boundaries, while barriers are created by water laid sands, gravels, or cobble. Barriers may be simple and continuous where the shoreline is straight or intermittently formed through multiple cycles of deposition and erosion. Deltas and persistent bars may also act as partial barriers in broad embayments where adjacent shorelines provide additional protection. A comprehensive discussion of barriers and lagoon systems is given by Jaworski et al. (1979).

Barriers must be sufficiently stable to reduce wave energy and erosion to the degree that vegetation can become rooted and sediments accumulate. Lake level influence is expressed through permanent or intermittent connecting channels or underground seepage. Water levels are usually augmented by inputs from tributary streams, and the hydrologic connection between lake and lagoon is more permanent where tributary flows are high. A pond or shallow, open water area is often associated with these systems.

The full wetlands continuum can develop in conjunction with barrier/lagoon systems. Habitat complexity and community composition are related to the interplay of topographic conditions and water regimes within the lagoon. Broad expanses of robust and meadow emergent communities dominate these systems throughout the region.

Embayed Wetlands. Upland peninsulas formed by bedrock outcrops or resistant soil materials provide protection for shallow water areas cut into the shoreline of the mainland or large islands. The wetlands which develop in these embayed or enclosed positions vary widely from thin vegetated ribbons, where the landward extent is shallow, to broad and complex systems, where bays are deeply cut. Tributary streams may flow into the lake basin through larger bays, and organic and mineral sediments derived from adjacent uplands may accumulate.

Since the mouth of the bay lacks a protective barrier, open water areas can be extensive. Embayed wetlands usually include greater topographic and substrate variability than barrier/lagoon systems, resulting in greater habitat diversity.

Streamside Wetlands. Riparian wetlands extend inland along the floodplains and banks of tributary streams entering the lake basin. Their extent is a function of floodplain width, being greatest along larger streams with broad floodplains and least where stream banks are steep. Since most tributary streams enter the lakes through lagoons and bays, the distinction between streamside wetlands and those of embayed or barrier/lagoon systems is imperfect. Distinctions are present in hydrology, because of the influence of upstream sources ón floodplain hydrographs, and in vegetation patterns, with more discrete lateral patterning of communities occurring along topographic gradients which run generally perpendicular to stream channels.

Island and Shoal Systems. As islands and shoals rise to the water surface, they create shallow water sites which provide minimal protection from wind and waves. Components of the wetlands continuum may develop in these high energy sites through the successful colonization by rooted macrophytic species. Emergent forms are usually prohibited by physical forces and the instability of sediments. The submerged aquatic plant communities which develop in these locations are compositionally related to those of adjacent, more protected locations. They are enormously important to fish and wildlife species during the fall and winter when detritus accumulates in their shallows.

The Hydrologic Regime and Community Development

We studied the relationships between environmental variables and wetland community composition by sampling net primary production at intervals over a growing season in plots located in two wetland systems along the Lake Ontario shoreline (Gilman, 1976). Peak standing crop biomass was used as a measure of species importance. The water regime was characterized, and a comprehensive series of physical and chemical substrate variables were determined simultaneously.

Four measures of water regime (mean, maximum, and minimum water levels and drawdown) were significantly correlated with species composition. Additional strong correlations were demonstrated between substrate features (organic matter content, exchangeable bases and textural variables in shallow and deep sediment layers) and community composition, and between substrate features and water regime variables. We interpreted these results to mean that the water regime is the overriding environmental factor which regulates the occurrence of wetland communities and modifies the relative significance of other variables. The correlation of substrate variables with community composition describes a tendency for established wetland vegetation to alter its environment through the deposition of organic matter and the entrapment of sediments. However, we interpreted this tendency to be depth related and dependent upon the primary influence of the hydrologic regime in patterning community composition. The significance of water regime variables in the wetland environment complex has been similarly demonstrated in other systems (see Quennerstedt, 1956; Dirschl and Coupland, 1972; Millar, 1973).

The importance of the water regime in community composition and development is emphasized by the successful construction of direct gradient models to summarize these studies. This one-dimensional ordination technique is only applicable where a strong overriding complex variable is responsible for community organization (Whittaker, 1967). Figure 1 illustrates progressive changes in species composition along a mean annual water depth gradient in Campbell Marsh, a riparian wetland system in Jefferson County, New York. Importance values of the major plant species are plotted over the position of each sample station along a linear gradient of mean annual water levels. Individual species importance curves were smoothed by calculating a moving average in 10 cm depth classes and replotting the adjusted value over the midpoint of that class. Details of model construction are given in Gilman (1976).

The wetlands continuum presented in Figure 1 also provides insight into the response of plant species to modifications in the prevailing environmental complex. The tolerance range or niche width (McNaughton and Wolf, 1970) of each dominant species is illustrated over the segment of the water depth gradient which corresponds to its distribution. For example, Typha glauca (tg) occupies a niche characterized by mean annual water depths from 6 to 56 cm, and its niche overlaps with that of Chara vulgaris (cv) at the deep water extreme. Thus, high water conditions which consistently exceed the tolerances of an established cattail community at its deep water edge would be predicted to result in a new community dominated by Chara vulgaris (cv), Najas flexilis (nf), Polygonum coccinium (pc) and Ceratophyllum demersum (cd) in wetlands where this model would apply.

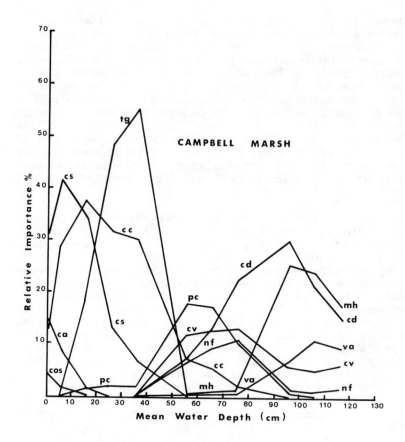

Figure 1. Distribution of wetland plant species along a water depth gradient at Campbell Marsh. Species indicated are <u>Myriophyllum heterophyllum</u> mh, <u>Ceratophyllum demersum</u> cd, <u>Vallisneria americana</u> va, <u>Chara vulgaris</u> cv, <u>Najas flexilis</u> nf, <u>Polygonum coccineum</u> pc, <u>Calamagrostis canadensis</u> cc, <u>Typha glauca</u> tg, <u>Carex stricta</u> cs, <u>Cornus amomum</u> ca, <u>Cornus stolonifera</u> cos (from Gilman, 1976).

FACTORS INFLUENCING WETLAND COMPOSITIONAL STABILITY

Wetland systems are simultaneously robust and fragile. They prosper at the edge of land/water interfaces which are subjected to environmental conditions of substantial annual variability. Yet their intimate linkage with the hydrologic regime renders them sensitive to change when modest directional shifts in hydrologic variables occur.

Water Levels and Compositional Change

Numerous studies have demonstrated that coastal wetlands are intimately tuned to their water regime and that compositional changes closely follow modifications in that regime (Gosselink and Turner, 1978; Weller, 1978). Jaworski et al. (1979) studied representative Great Lakes wetlands using a combination of field survey and historical reconstruction based on aerial photographs. They concluded that the total area of coastal wetlands has oscillated over recent hydroperiods characterized by high and low water conditions. High water level periods give rise to destructive action of waves on protective shoreline features, such as bars, deltas, and barriers, and on those wetlands which have developed due to their protective influence. More gradual shifts in the internal composition of established wetlands occur in response to episodes of both high or low water. Such changes can be predicted by "plant community displacement models" which display the elevational distribution of plant communities in relation to actual or inferred water levels.

Studies of wetland composition along the Lake Ontario shoreline in 1974 (Geis and Kee, 1977) occurred simultaneously with a major episode of high water die-off (see Geis, 1979). A total of 10.7 percent of the wetland area in a single county contained dead vegetation, and dead cover types were delineated in communities dominated by emergents, shrubs and trees. Emergent communities experienced the greatest impact, with die-off occurring most commonly at the leading edge of cattail communities and sedge-grass meadows. A recent study by Bush and Lewis (1984) described multiple cycles of wetland community expansion and contraction along the same shoreline through analysis of historical aerial photographs.

Although the precise sequence of events precipitating 1974 wetland die-off was not established, analysis of water level data suggested that higher than normal winter water levels were responsible. The episode was similar in many respects to the report by McDonald (1955) of extensive die-off of emergent wetlands at Point Mouillee, Michigan, on Lake Erie which corresponded with abnormally high winter lake levels.

Wetlands in Winter

The progressive development and dissipation of ice and snow along the Great Lakes shoreline creates environmental

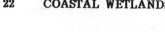

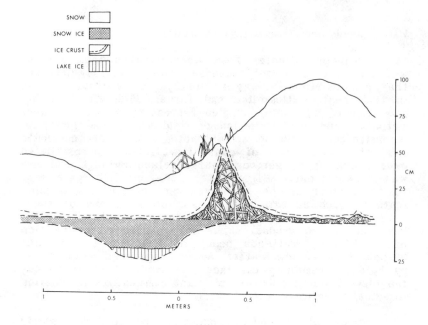

Figure 2. Generalized diagram showing snow and ice conditions in shoreline wetlands along the St. Lawrence River during the winter of 1978–79 (from Geis, Ruta and Hyduke, 1985).

conditions and system sensitivities which differ markedly from those during the ice out period of the year. These conditions can have a profound influence on wetland stability and internal species composition. Studies conducted along the Lake Ontario and St. Lawrence River shorelines (Geis, Ruta and Hyduke, 1985) have identified four groups of processes which are of particular significance: the formation of unique seasonal habitats beneath the wetland snowpack; the freeze-down and dewatering of otherwise saturated bottom deposits; the incorporation of bottom material into the ice sheet such that disruption and benthic turnover occur during spring break-up; and the creation of physical linkage between shoreline substrate and the ice sheet such that edge breakage may occur when the ice sheet moves out. All four processes are intimately linked to the prevailing water regime, and winter water levels may substantially modify their propagation, development and impact.

     Seasonal Habitats Beneath Wetland Snowpack. A generalized model of wetland snow and ice conditions is given in Figure 2. Deep snow covers accumulate in wetlands from direct snowfall and the redeposition of wind-blown snows off

adjacent bays and channels. Snow depths within the wetland
are highly variable. Depth is greatest in drift lines
which accumulate at the interface between the wetland
and adjacent bay ice sheet. Within the wetland itself,
depth is greatest in drifts associated with clumps of
dead standing vegetation separated by hollows where the
vegetation became flattened. Ice below the snowpack is
also irregular in distribution and variable in thickness.
Where present, it may be frozen to the bottom, floating
or suspended due to water level fluctuations. The deep
snowpack insulates the wetland surface and prevents ice
formation, especially in areas where a loose layer of
dead vegetation remains. This condition may change in
the spring when thawing and refreezing create a snow ice
cover of short duration.

The wetland snowpack is low in density, and much
of it consists of recrystallized snows resulting from
the migration upward of water vapor from lower, warmer
snows resting on unfrozen wetland substrates. The redistribution
of water vapor to upper, colder portions of the snowpack
results in the formation of voids of irregular size and
shape near the wetland surface. Voids are prevented from
collapse by vegetative clumps and the formation of ice
crusts on their inner surfaces. Chimney-like voids extend
through the snowpack around clumps of dead vegetation.
They channel warm water vapor from the unfrozen surface
through the snowpack and out the chimneys. Voids and
chimneys form an interconnected matrix of protected habitats
beneath the snowpack. Similar features have been described
by Marshall (1978).

Winter water levels may influence the wetland snowpack
in several ways. Lower water levels and subsequent draining
from the wetland during early winter may increase the
degree and distribution of frozen substrate, especially
if freezing conditions precede significant snow accumulation.
Higher water levels following snow deposition may flood
these habitats, causing slush formation and subsequent
freezing of the wetland surface. Either condition may
significantly alter the wetland snowpack, freezing benthic
deposits and adversely modifying overwintering habitats.

Freeze-Down of Sediments at the Wetland Edge. The
general sequence of shoreline ice formation at wetland
interfaces along the St. Lawrence River is illustrated
in Figure 3. Although the process occurs continuously
and is subject to variations in local shoreline conditions,
three general phases can be recognized: first freeze-
down, shown as the upper portion of Figure 3; maximum

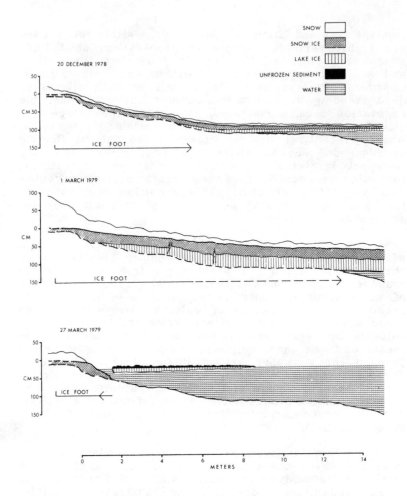

Figure 3.   Three phases of shoreline ice formation at wetland interfaces along the St. Lawrence River during the winter of 1978-79.   December 20, 1978, corresponds to first freeze-down; March 1, 1979, corresponds to maximum extent of ice; and March 27, 1979, corresponds to conditions at spring break-up (from Geis, Ruta and Hyduke, 1985).

extent of ice, shown as the middle portion of Figure 3; and spring break-up, shown as the lower portion of Figure 3.

The linkage of wetlands to the ice cover of adjacent shallow water areas occurs through the development of a feature called the ice foot (Marshall, 1978).   The ice foot is defined here as a zone of grounded ice which incorporates frozen sediments and extends into the wetland edge to a variable extent. It forms initially by direct freezing of substrates lying above the prevailing water level at the time of first ice formation or by the deposition and refreezing of slush, sludge and new ice blown ashore at the wetland edge prior to the formation of a stable ice cover. As the ice cover builds and thickens, the depth of freeze-down increases and the ice foot migrates outward.

Between first freeze-down and maximum extent of ice, the zone of grounded ice migrates progressively outward as a result of increasing water levels and the thickening of the ice pack through snow ice and lake ice formation. Snow ice forms through the capillary seepage of water through the ice into snows falling on the surface, while lake ice forms by direct freezing of water at the lower ice surface. Snow ice formation is also enhanced by bleeding upward of water at hinge cracks. Maximum extent of ice, illustrated in the middle portion of Figure 3, is characterized by a broad ice foot zone extending outward from the wetland edge.

At the onset of spring break-up, melting begins in the stream channels and ice erosion progresses toward the wetland edge. In the lower portion of Figure 3, ice erosion has progressed well within the ice foot to the point where grounded ice remained attached to the wetland edge. The outer ice foot remains anchored. The sediment in the outer zone has been lifted from the bottom and is shown floating on the last remnants of the ice sheet. Spring water level increases generally precede final ice erosion at the shoreline and, consequently, the outer edge of the anchored ice becomes submerged. This phenomenon is called moating and is characteristic of ice sheet erosion at most shoreline contacts. Depending on the timing and degree of water level change, erosion of the edge proceeds gradually by melting out, or abruptly by lifting and dispersal of ice, sediment and vegetation.

Incorporation of Bottom Material into the Ice Pack. A zone of grounded ice linking the shoreline to the ice pack and incorporating frozen sediments occurs consistently along the St. Lawrence River. It may extend 40 to 50 meters outward from wetlands in protected bays and 5 to 10 meters outward along rocky shorelines where bottom gradients are steeper. This ice foot zone forms initially through direct freezing of exposed bottoms and later through the accumulation of ice and snow as the winter progresses.

Sediments may become incorporated within the ice pack by direct freezing, by bottom lifting and refreezing as water levels increase later in the winter, or by progressive thickening and grounding of the ice sheet throughout the winter. All portions of the ice foot zone experience some benthic disruption due to the lifting of sediments incorporated into the ice sheet. However, this natural disturbance is most severe within the first-formed ice foot where the frozen layer is deepest and lifting during spring break-up is most pronounced. Within this zone, overwintering plant parts are displaced and benthic materials are dislodged and redistributed.

Winter water levels exert a profound influence on the formation and impact of the ice foot zone. Lower water levels in early winter result in greater sediment exposure and a more extensive first-formed ice foot. Higher water levels in midwinter cause more extensive bottom lifting and refreezing in the first-formed ice foot. These same conditions also cause thicker ice sheets inshore, as a result of the bleeding of water through hinge cracks. Higher water levels during spring break-up cause more extensive bottom lifting, especially if water level rise precedes bottom melting in the outer portions of the developmental ice foot.

Linkage of Ice Sheet to the Wetland Edge. Physical linkage between the ice sheet and the frozen wetland or shoreline edge creates a tendency for portions of the shoreline to become broken off as the ice sheet moves outward during spring break-up. This process can be most disruptive at wetland interfaces during the high water years when ice extends some distance into the wetland. High spring water levels followed by refreezing may also create these conditions.

Effects of Ice Conditions on Vegetation at the Shoreline. Freeze-down, bottom lifting and the disruption of benthic deposits within the ice foot zone result in pronounced differences in standing crop biomass and species composition of shallow water littoral plant communitites. These differences are more pronounced at wetland interfaces within bays than at shoreline stations along the open channel. We collected standing crop biomass samples twice (June and August) at sample points established in relation to the maximum extent of ice foot at a series of transects along the St. Lawrence River in 1979. All vegetation was clipped by a diver usng a one meter square sample frame.

Mean standing crop biomass at wetland margins in June averaged 2.3 times greater outside the ice foot zone than at comparable stations within the zone. Species richness was also reduced within the ice foot zone, as was the presence of overwintering biomass. These trends were maintained in August-September samples, in which mean standing crop biomass outside the ice foot zone exceeded that within the zone by an average of 2.1 times.

Stability and Change in Coastal Wetlands

The diversity of wetland communities which occur in close proximity, often in adjacent bands or zones, gives an initial impression of structure and order. In some systems, zones of similar growth form are arrayed progressively along transects of decreasing depths of sediment and organic materials. The temptation is to infer a hydrosere of community advancement through substrate building and amelioration, even though that model has been demonstrated to be inadequate (Mandossian and McIntosh, 1960; Auclair et al., 1973; Geis, 1979).

Coastal wetland systems do exhibit autogenic substrate accumulation and modification, but as stated by Auclair et al. (1973), these tendencies "may be easily outweighed by turbulence in the water and displacement or deposition of sediments by water current and wave action." The pulses of community die-off related to barrier breakage; the erosion of protective shoreline features or cyclic water level modifications; the redistribution of sediments within systems; and the disruption of benthic materials and perennating plant parts by ice-propagated disruptions in winter are evidence of a continuous, allogenic environmental dimension which opposes these tendencies. Reference to both autogenic and allogenic tendencies is necessary for a comprehensive assessment of the dynamic or successional relationship of coastal wetland systems.

DISCUSSION

Ted R. Batterson
Department of Fisheries and Wildlife
Michigan State University
East Lansing, Michigan 48824

Overall, Geis' article is good, and he presents a fairly thorough treatment of the environmental variables which affect the composition and distribution of Great Lakes wetlands. I agree with his inclusion of submersed

macrophyte communities into the "wetlands continuum" and
would suggest that the U.S. Fish and Wildlife Service
(Cowardin et al., 1979) incorporate submersed wetland
as one of their habitat classes.

Geis presents four broad categories of wetland systems.
From his description of these categories, I am not sure
under which one wetlands of western Lake Erie would be
placed. I also wonder how wetlands of the connecting
channels (e.g., St. Marys River) would be categorized.
They are streamside wetlands, but Geis' definition would
need to be modified if these wetlands are to be included
in this category.

He devotes the last half of the article to the influences
winter and freezing water conditions have on wetlands.
To date, there has been a paucity of published data in
regard to this matter. More studies need to be conducted
which would document the effects that ice-on, stable ice
conditions, or ice-off periods might have on wetlands;
ice-off being the time at which the ice is dissipating
or receding as opposed to the ice-free period of the year.

Our work over the last several years on the wetlands
of the St. Marys River support Geis' contention that winter
and freezing conditions can have a profound effect upon
these plant communities. An understanding of what affects
the stability and species composition of wetlands of the
St. Marys River is important since they are the dominant
primary producers of this system. On a relative basis,
the emergent wetlands are 236 times more productive and
submersed wetlands 16 times more productive than the phytoplankton
of the river (Liston et al., 1985). Maps from the river
show open areas in the emergent vegetation community commonly
occur in shallow mid-portions of these wetlands (Liston
et al., 1985). Borings taken during the winter of 1983
along transects perpendicular to the shore in these wetlands
indicated that an ice foot extended up to 25 cm into the
hydrosoil. Over 95 percent of the underground biomass
is located within the upper 15 cm of sediment, and it
is these tissues which function as the perennating source
for the following year's growth. Ice movement which could
disrupt or displace these over-wintering tissues would
impact the emergent wetlands of this system. This is
of major importance to the overall structure and function
of these communities since we have found that the plants
slowly recolonize openings that develop within the wetland
boundaries (Liston et al., 1985), and certain species
are more sensitive to perturbations than others. We have
hypothesized that one of the agents causing disruption
of emergent vegetation along the St. Marys River has been

due to ice movement, either as a result of natural processes or caused by the passage of commercial vessels.

LITERATURE CITED

Auclair, A.N., A. Bouchard, and J. Pajaczkowski. 1973. Plant composition and species relationships on the Huntington Marsh, Quebec. Can. J. Bot. 51:1231-1247.

Bush, W.D.N. and L.M. Lewis. 1984. Responses of wetland vegetation to water level variations in Lake Ontario. Pp. 519-524 in: Proc. Third Conf. Lake and Reservoir Management. U.S. Environ. Protection Agency, Washington, DC.

Cowardin, L.M., V. Carter, F.C. Golet, and E.T. LaRoe. 1979. Classification of wetlands and deep water habitats of the United States. FWS/OBS-79/31. Office of Biological Services, Fish and Wildlife Service, U.S. Department of the Interior, Washington, DC. 103 pp.

Dirschl, H.J. and R.T. Coupland. 1972. Vegetation patterns and site relationships in the Saskatchewan River delta. Can. J. Bot. 50:647-676.

Geis, J.W. 1979. Shoreline processes affecting the distribution of wetland habitat. Trans. N. Amer. Wildl. Nat. Resour. Conf. 44:529-542.

Geis, J.W. and J.L. Kee. 1977. Coastal wetlands along Lake Ontario and the St. Lawrence River in Jefferson County, New York. SUNY College Environ. Sci. For., Syracuse, NY. 130 pp.

Geis, J.W., P.J. Ruta, and N.P. Hyduke. 1985. Wetland and littoral zone features of the St. Lawrence River in winter. Unpublished manuscript.

Gilman, B.A. 1976. Wetland communities along the eastern shoreline of Lake Ontario. MS. Thesis. SUNY College Environ. Sci. For., Syracuse, NY. 187 pp.

Gosselink, J.G. and R.E. Turner. 1978. The role of hydrology in freshwater wetland ecosystems. Pp. 63-78 in: R.E. Good, D.F. Whigham and R.L. Simpson (eds.), Freshwater Wetlands. Ecological Processes and Management Potential. Academic Press, New York.

Jaworski, E., C.N. Raphael, P.J. Mansfield, and B.B. Williamson. 1979. Impact of Great Lakes water level fluctuations on coastal wetlands. USDI Office Water Resources Technology. Contract Report 14-0001-7163. 351 pp.

Liston, C.R., C.D. McNabb, D.Brazo, J. Bohr, J.R. Craig, W.G. Duffy, G. Fleischer, G.W. Knoecklein, F. Koehler, R. Ligman, R. O'Neal, P. Roettger, and M. Siami. In Press. Environmental baseline studies of the St. Marys River during 1982 and 1983 prior to proposed extension of the navigation season. Project Report. Department of Fisheries and Wildlife, Michigan State University, East Lansing, Michigan. Contract No. 14-16-0009-79-013. Submitted to the U.S. Fish and Wildlife Service, Fort Snelling, Minnesota.

Mandossian, A. and R.P. McIntosh. 1960. Vegetative zonation on the shore of a small lake. Amer. Midl. Natur. 64(2):301-308.

Marshall, E.W. 1978. Ice survey studies related to demonstration activities. Tech. Rep. H, pp. H1-H92 in Environmental Assessment FY 1979 Winter Navigation Demonstration on the St. Lawrence River, Vol. 2. SUNY College Environ. Sci. For., Syracuse, NY.

McDonald. M.E. 1955. Causes and effects of a die-off of emergent vegetation. J. Wildl. Manage. 19:24-35.

McNaughton, S.J. and L.L. Wolf. 1970. Dominance and niche in ecological systems. Science. 167:131-139.

Millar, J.B. 1973. Vegetational changes in shallow marshes under improving moisture regime. Can. J. Bot. 51:1443-1457.

Quennerstedt, N. 1956. The effect of water level fluctuations on lake vegetation. Proc. Intern. Assoc. Theor. Appl. Limno. 13:901-906.

Ruta, P.J. 1981. Littoral macrophyte communities of the St. Lawrence River, New York. MS. Thesis. SUNY College Environ. Sci. For., Syracuse, NY. 153 pp.

Sculthorpe, C.D. 1967. Biology of aquatic vascular plants. Edward and Arnold Ltd., London. 610 pp.

van der Valk, A.G. and L.C. Bliss. 1971. Hydrarch succession and net primary production of oxbow lakes in central Alberta. Can. J. Bot. 49:1177-1199.

Weller, M.W. 1978. Management of freshwater marshes for wildlife. Pp. 267-284 in: R.E. Good, D.F. Whigham, and R.L. Simpson (eds.), Freshwater Wetlands. Ecological

Processes and Management Potential.   Academic Press, New York.

Whittaker, R.H.   1967.   Gradient analysis of vegetation. Biol. Rev.   42:207-264.

CHAPTER 3

VEGETATION DYNAMICS, BURIED SEEDS, AND WATER LEVEL FLUCTUATIONS ON THE SHORELINES OF THE GREAT LAKES

P. A. Keddy
Department of Biology
University of Ottawa
Ottawa, Ontario, Canada K1N 6N5

A. A. Reznicek
University of Michigan Herbarium
Ann Arbor, Michigan 48109

"This rise and fall of Walden [Pond] at long intervals serves this use at least; the water standing at this great height for a year or more, though it makes it difficult to walk round it, kills the shrubs and trees which have sprung up about its edge since the last rise....and, falling again, leaves an unobstructed shore; for, unlike many ponds and all waters which are subject to a daily tide, its shore is cleanest when the water is lowest....By this fluctuation the pond asserts its title to a shore, and thus the shore is shorn, and the trees cannot hold it by right of possession."

H. D. Thoreau, 1854.

INTRODUCTION

Since the above observation was made, there has been little improvement in our understanding of shoreline vegetation and its dynamic interaction with water levels. Our objectives were to review what is known about water levels and shoreline vegetation in the Great Lakes, and produce a qualitative model to describe the effects of fluctuating water levels. We soon found that few published

papers provide quantitative data on changes in lakeshore
vegetation with water level, and fewer still discuss the
Great lakes specifically. We therefore reviewed what
is known about the effects of water level fluctuations
on vegetation in other North American wetlands and sought
some general principles which could likely be applied
to Great Lakes shorelines. These were supplemented with
some descriptive papers on the flora of shoreline marshes
(e.g., Dore and Gillett, 1955; Hayes, 1964; Stuckey, 1975;
Bristow et al., 1977; Jaworski et al., 1979; Fahselt and
Maun, 1980) to provide at least some qualitative observations
on the effect of water level fluctuations on species composition.
In the first part of this paper, we will discuss the dynamics
of shoreline vegetation with changing water levels and
present our qualitative model. In the second part, we
will discuss the flora associated with different vegetation
types.

PART I: WATER LEVEL FLUCTUATIONS AND VEGETATION

Marsh vegetation dynamics have received much more
attention (e.g., Harris and Marshall, 1963; van der Valk
and Davis, 1978, 1979; van der Valk, 1981; Smith and Kadlec,
1983) than lakeshore vegetation dynamics (Keddy and Reznicek,
1982; Nicholson and Keddy, 1983). We have therefore had
to extrapolate from the former as well as the latter.
Lakeshore vegetation may be little different from a typical
marsh when it occurs in sheltered, gently sloping bays.
The steeper the shoreline becomes, however, the more evident
zonation becomes and the narrower the wetland becomes.
The factors influencing lakeshore zonation have recently
been reviewed by Hutchinson (1975) and Spence (1982).
Although introductory ecology texts still sometimes infer
that this zonation represents succession, in most cases
the two phenomena are unrelated on lakeshores. Zonation
is better viewed as simply the response of different wetlands
species to fluctuating water levels. Lakeshore wetlands
are also often exposed to erosion from water. In such
cases, the upper shorelines are eroded and deposition
occurs in the deeper water. Bernatowicz and Zachwieja
(1966) have distinguished ten types of littoral zones,
considering primarily the effects of erosion on different
substrate types. As well as being influenced by water
depth, lakeshore wetlands may also have strong gradients
parallel to the waterline as waves sort material from
highly exposed shores to sheltered shores (Hutchinson,
1975; Spence, 1982; Davidson-Arnott and Pollard, 1980;
Keddy, 1982, 1984).

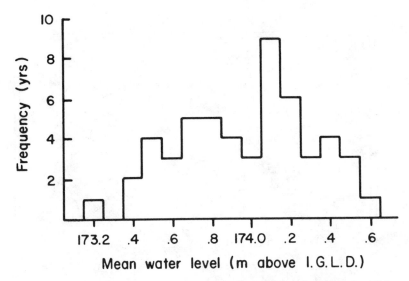

Figure 1. Year to year variation in August mean monthly water level in Lake Erie, Port Stanley, 1927-1980 (except 1978).

Since seasonal (or within-year) water level fluctuations are superimposed upon long-term (or among-year) fluctuations, we will first consider the effects of long-term fluctuations. As a typical example of water level changes, Figure 1 shows the August water levels at Port Stanley on Lake Erie over a 53-year period. The present distribution and abundance of shoreline species will be determined by past as well as present water levels. How far into the past, or with what weighting, we do not know. The extreme highs and lows will produce the most rapid vegetation change; we will consider low periods first.

Low Water Periods

During low water periods, several changes can be expected. Soil chemistry may change dramatically as soils become less anoxic (Ponnamperuma, 1972). Some plant species will change their growth form to accommodate drier conditions (Sculthorpe, 1967; Hutchinson, 1975), but the vegetation will usually change dramatically as species intolerant of drying die and are replaced by species emerging from reserves of buried seeds. Much emphasis has been placed on documenting this regeneration from buried seeds (e.g., Kadlec, 1962; Harris and Marshall, 1963; Salisbury, 1970;

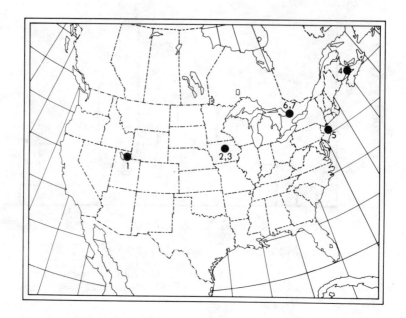

Figure 2. Location of seed bank studies used in compiling Tables 1 and 2. Numbers refer to those in Table 1.

van der Valk and Davis, 1976, 1978, 1979; van der Valk, 1981; Keddy and Reznicek, 1982; Smith and Kadlec, 1983). Table 1 shows the density of seeds in soil samples from different North American wetlands (see Figure 2 for the locations of the wetlands).

Several patterns are suggested by the data in Table 1. While densities of buried seeds are high in most wetland vegetation types, they appear to increase from prairie marshes to coastal marshes to lakeshores. The only lake studied (Matchedash Lake) has a very rich flora on gently-sloping sand and gravel shorelines. Since low water periods allow many species to replenish reserves of buried seeds, the high densities of buried seeds might be attributed to frequent low water levels in Matchedash Lake. Given the regular occurrence of low water periods in the Great Lakes, comparable data would be most interesting.

The data in Table 1 also suggest that seedling densities are highest in the littoral zone. Perhaps these intermediate

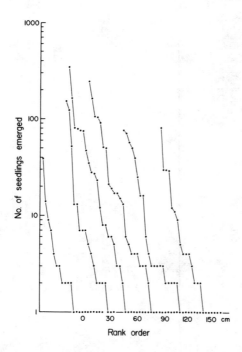

Figure 3. Relative abundances of seedlings from samples representing six depths on the shoreline of Matchedash Lake. 0 cm marks the water depth when the samples were taken, but falling water levels could expose all depths.

depths have the appropriate combination of submersed and emersed periods for seeds to accumulate.

Many genera found in seed banks are common on the shorelines of the Great Lakes. Table 2 lists genera which could be expected to occur in samples from Great Lakes marshes.

Although the densities of buried seeds may be high, samples are often dominated by only a few species. Figure 3 shows the relative abundances of seedlings from six water depths on the shoreline of Matchedash Lake. The dominant species at all depths was Hypericum majus, except for 0 cm where it was Panicum implicatum. Many species were present as less than five individuals.

Moore and Wein (in press) point out that accumulation of organic matter may gradually remove nutrients from

Table 1. Comparison of freshwater wetland seed banks, using data from moist (not flooded) germination treatments. Van der Valk and Davis (1979) was excluded because flooded and moist conditions could not be separated in their tables; Haag (1981) was excluded because only flooded conditions were used.

| Study | Sample Unit | | | Site | Seedlings m⁻² |
	Surface area (cm²)	Depth (cm)	Number		
PRAIRIE MARSHES					
1. Smith and Kadlec, 1983[1]	400	4	25	Typha spp.	2682
				Scirpus acutus	6536
				S. maritimus	2194
				Phragmites australis	2398
				Distichlis spicata	850
				Open water	70
2. van der Valk and Davis, 1978[2]	~225	~4-5	12	Open water	3549
				Scirpus validus	7246
				Sparganium eurycarpum	2175
				Typha glauca	5447
				Scirpus fluviatilis	2247
				Carex spp.	3254
3. van der Valk and Davis, 1976[2]	~230	~4-5[4]	27	Open water	2900
			15	Typha glauca	3016

					Scirpus fluviatilis
		3	10		319
FRESHWATER COASTAL MARSHES					
4. Moore and Wein, in press[5]	125	12	10	Typha latifolia	14768
				Former hay field	7232
				Myrica gale	4496
5. Leck and Graveline, 1979[6]	100	10	10	Streambank	11295
				Mixed annuals	6405
				Ambrosia trifida	9810
				Typha latifolia	13670
				Zizania aquatica	12955
LAKESHORE					
6. Nicholson and Keddy, 1983[7]	7.5	10	49	Lakeshore, 75 cm water	38259
7. Keddy and Reznicek, 1982[8]	7	5	75	Waterline	1862
				30 cm below water line	7543
				60 cm below water line	19798
				90 cm below water line	18696
				120 cm below water line	7467
				150 cm below water line	5168

[1] Table 1
[2] Table 2
[3] Table 1, using highest density for each site type we multiplied by 14.5 to obtain no./m^2
[4] based on van der Valk and Davis, 1978
[5] Table 1, spring samples
[6] data from text (p. 1009) which had been extrapolated from Table 1 to include the depths 2-4 and 6-8 cm.
[7] Table 1, 0 to 10 cm depths only used to avoid gaps in core beyond this depth
[8] recalculated from original data

Table 2. Genera with widespread occurrence in wetland seedbanks[1], based on species germinating during moist treatments (i.e., germinating during conditions simulating low water periods).

Genus	Prairie Marshes[2]	Freshwater Coastal Marshes[3]	Lakeshores[4]
Bidens	*	*	*
Carex	*		*
Cuscuta	*	*	
Cyperus	*	*	
Echinochloa	*	*	
Galium	*	*	
Hypericum		*	*
Impatiens	*	*	
Juncus		*	*
Leersia	*		*
Ludwigia		*	*
Lycopus	*		*
Panicum	*	*	*
Polygonum	*	*	*
Sagittaria	*	*	
Scirpus	*	*	*
Spirea		*	*
Typha	*	*	
Viola		*	*

[1]80 genera were recorded in these studies. Many were restricted to a single study (43,*) or a single vegetation type (18). Only widespread genera occurring in at least two vegetation types are listed here. Other genera recorded were:

prairie marshes: Alisma, Amaranthus, Asclepias*, Berula*, Brachyactis*, Chenopodium, Cicuta*, Cirsium*, Distichlis*, Eleocharis, Epilobium*, Eragrostis*, Glyceria*, Iris*, Mentha*, Mimulus*, Penthorum*, Phragmites*, Polypogon*, Populus*, Potamogeton*, Ranunculus*, Rorippa, Rumex, Salicornia*, Scutellaria*, Sium, Solanum*, Sparganium, Stachys*, Urtica;
freshwater coastal marshes: Acnida*, Ambrosia*, Callitriche*, Gratiola*, Hibiscus*, Lythrum*, Mikania*, Peltandra*, Pilea*, Potentilla, Veronica*, Zizania*;

marsh vegetation. They suggest that fire can be used during low water periods to remove accumulated organic matter and "rejuvenate" the marsh. Like Leck and Graveline (1979) and van der Valk and Davis (1979), they found that viable seeds were relatively common at depths of 12 cm or more and concluded that even after a very severe burn, buried seeds would allow revegetation. Some of their conclusions may not apply to lakeshores. First, the rate of accumulation of organic matter on lakeshores may be much lower, since it can oxidize during low water periods, or be eroded and exported during high water periods. If the organic layer is shallow, there may be only a shallow layer of buried seeds. The only data from a lakeshore (Nicholson and Keddy, 1983) show that 81 percent of buried seeds occurred in the top 2 cm. Disturbance from fire could therefore have much greater effects on lakeshores.

High Water Levels

Rising water levels will change soils from oxic to anoxic (Ponnamperuma, 1972). Organic matter and fine particles (e.g., silt and clay) may be removed by water circulation (Jaworski et al., 1979). Simultaneously, mud flat species disappear (e.g., Salisbury, 1970; van der Valk, 1981). Emergent species will propagate vegetatively under shallow water, but will gradually die out under deeper water (Harris and Marshall, 1963; van der Valk and Davis, 1978). Farney and Bookhout (1982) describe how high water levels in Lake Erie converted emergent vegetation to open water. Even cattails (Typha spp.), which covered more than 20 percent of their study area, were eliminated. Other common cover types such as Hibiscus palustris and Leersia oryzoides, also disappeared. Jaworski et al. (1979) provide many similar examples from Lakes Michigan, Huron, St. Clair and Erie. High water periods therefore eliminate one group of marsh species and allow them to be temporarily replaced by floating-leaved and submerged species more tolerant of flooding. The causes of death of emergents are unclear. Some species which

---

(Table 1 footnotes continued)

lakeshore: Agrostis, Calamagrostis*, Chamaedaphne, Cladium, Drosera, Dulichium*, Eriocaulon, Gnaphalium*, Linum*, Lysimachia*, Muhlenbergia, Myrica*, Rhexia, Rhynochospora, Rubus*, Solidago*, Triadenum*, Xyris*.
[2] Smith and Kadlec (1983); van der Valk and Davis (1978, 1976).
[3] Moore and Wein (in press) genera list may not be complete; Leck and Graveline (1979).
[4] Nicholson and Keddy (1983); Keddy and Reznicek (1982).

are intolerant of flooding produce toxic ethanol during
anoxic conditions; species more tolerant of flooding possess
an alternate pathway which leads to the accumulation of
the far less toxic malate (McManmon and Crawford, 1971;
Barclay and Crawford, 1982). In prairie marshes, muskrat
damage and disease may also contribute to the death of
emergents during high water periods (van der Valk and
Davis, 1978).

High water levels have a second important effect
on lakeshore marshes: the elimination of trees and shrubs.
Recently, we found evidence of this process in Matchedash
Lake (Keddy and Reznicek, 1982). The upper limit of many
herbaceous species on lakeshores coincides with the lower
limits of woody species; and where waves or ice remove
shrubs, herbaceous species grow further landward (Keddy,
1983). If woody plants set the upper limit of herbaceous
species, then high water levels, by eliminating woody
plants, may increase the area occupied by herbaceous wetlands
species. An observation consistent with this proposal
is that in small lakes with stable water levels, the shrub
zone frequently occurs right to the water line, leaving
only a narrow zone of emergents.

Several studies have erroneously implied and/or
concluded that high water levels have a negative impact
on shoreline vegetation, based simply on the observation
that "marsh" area temporarily decreases during high water
years (Jaworski et al., 1979; Lyon and Drobney, 1984;
also papers in this volume). In fact, high water levels
should eliminate woody plants (increasing marsh area)
and kill dominant species like Typha spp. (increasing
marsh diversity). To accurately assess the long-term
effects of occasional high water levels, one must collect
different data: first, we need to compare two groups
of marshes, one group with regular flooding in the past,
the other with stabilized water levels. Second, vegetation
data must be collected when both groups are in a low water
phase. We predict that occasional high water levels will
increase both the area and vegetation diversity of a marsh
during the following low water period.

Seasonal Fluctuations and Strand Vegetation

Water levels fluctuate on many time scales. Seasonal
fluctuations (Figure 4) are likely to have very different
effects from fluctuations with a period of a decade or
longer. In the latter case, population responses can
occur, with some species surviving only as buried seeds,

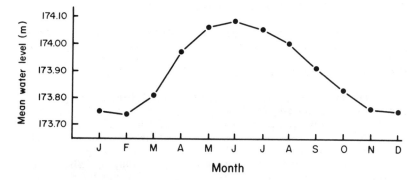

Figure 4. Seasonal (within-year) fluctuations in water level at Port Stanley on Lake Erie, averaging data from 1927-1980 (except 1978).

and others temporarily exploiting the existing conditions. With seasonal fluctuations, population responses are possible only for annuals which complete their life cycle rapidly. As the water level falls, different annuals will germinate and temporarily exploit favourable sites. In contrast, perennial species must be able to survive the entire range of conditions encountered during seasonal fluctuations in order to occupy a site during the growing season. Thus, they may produce different shoot morphologies as the season progresses (Sculthorpe, 1967; Hutchinson, 1975) and have different metabolic pathways for surviving anoxic periods (McManmon and Crawford, 1971; Barclay and Crawford, 1982). The annuals can escape seasonal fluctuations; the perennials must tolerate them. Seasonal fluctuations may increase species diversity. Stuckey (1975) observed at Put-in-Bay, Lake Erie, that "The greatest diversity of vegetation zones and greatest diversity of species within zones occur in that part of the marsh where the water level fluctuated the most throughout the season." He recognized 12 "dominant vegetation zones," seven of which were associated with fluctuating water levels.

At the very least seasonal fluctuations increase the annual component of the vegetation. For perennial species which can only germinate on exposed mud flats, the seasonal low may supplement or accentuate regeneration phases provided by long-term fluctuations. Lastly, since many wetlands species are apparently intolerant of continual submergence (e.g., Harris and Marshall, 1963; van der Valk and Davis, 1978), seasonal lows may allow shoreline species to occur deeper into the lake.

Water Level Fluctuations: A Natural Disturbance

Water level fluctuations are a natural form of disturbance. The role of natural disturbance in promoting vegetation diversity has been discussed by Grubb (1977), Connell (1978), Huston (1979), White (1979) and Grime (1979).

Disturbance has several quantifiable components including intensity and frequency. We do not yet know what intensity (amplitude) or frequency of disturbance from fluctuating water levels will maximize species diversity. Some combinations of high intensity and frequency have a negative impact on shoreline vegetation, as illustrated by the sparse vegetation of the margins of some hydroelectric reservoirs. Stabilizing water levels (reducing the intensity and frequency of disturbance) would also be expected to cause major changes in wetlands, particularly (1) the loss of species which regenerate during low water periods and (2) increased dominance by a few species such as woody plants and Typha spp.

A Model of Water Levels and Shoreline Vegetation in the Great Lakes

A model outlining the relationships of Great Lakes wetland vegetation types to water level fluctuations is depicted in Figure 5. This model is simplistic in that it considers only the role of water level fluctuations in determining wetland vegetation types. Topography, substrate type, wave action, latitude, water quality, fire, water currents, exposure, and length of time since the last high or low water phase are not considered. While the model thus is not refined enough to predict the occurrence of communities or species associations, we feel it is a useful conceptual framework for interpreting the large scale cyclic processes in Great Lakes wetlands.

The following discussion summarizes the hypothesized dynamics of the wetland types described earlier. Only strand vegetation is omitted, since it results primarily from various kinds of disturbance near the water line.

The upper part of the shore is dominated by woody species intolerant of flooding. They form forest and shrub thickets.

Wet meadow vegetation develops between the maximum high and present water level. The dynamics of this vegetation are probably similar to the dynamics of vegetation on

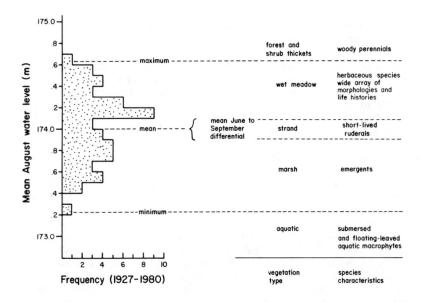

Figure 5. Proposed relationship between water levels and vegetation types on the Great Lakes shorelines. The water level data represent Lake Erie 1927-1980, minimum 1934, maximum 1973. The boundaries between the vegetation types will shift as water levels change; the strand, composed of short-lived ruderals, tracks the waterline with a width resulting from the fall in water levels from June to September. Other environmental factors such as slope, substrate type, wave action, water chemistry and fire will influence the species composition within each vegetation type.

shores of smaller lakes with fluctuating water levels (Keddy and Reznicek, 1982). During high water phases, these communities are narrow in width or even totally flooded. Woody plants that have invaded since the last high water level are killed, as are many herbaceous species. When water levels recede, wet meadow species re-establish from buried seeds and from individuals which survived on the upper fringes of the wet meadow zone. (We have used August water level data in Figure 5, but it is likely that higher water levels in June set the lower limits of woody plants; thus the upper limit of meadow species is probably higher than we have indicated.)

Between the present water line and the extreme minimum water level is the zone in which shallow marsh vegetation is best developed. The emergent aquatics can survive permanent flooding but many require occasional low water levels to expose the lake bottom in order for seedlings to establish. Thus, periodic seed recruitment of species can only occur above the extreme low water line, although some emergent aquatics can spread vegetatively into water deeper than the minimum low water line. A major difference between wet meadows and marshes is thus the relative frequency of flooding: meadows are occasionally flooded, whereas marshes are usually flooded.

Below the minimum low water level is the zone where aquatic vegetation survives continuously. In the shallower levels of the zone, emergent aquatics may invade during low water, although they will be eliminated again when water levels rise.

In the terminology of Jeglum et al. (1974), our forest and shrub thicket zone would include treed fen, thicket swamp and hardwood swamp. The wet meadow zone would include meadow marsh and graminoid fen, and the marsh zone would include deep marsh and shallow marsh. Jaworski et al. (1979) present some profiles which illustrate these zones in different wetlands.

An excellent starting point for future studies on water level fluctuations and shoreline vegetation would be van der Valk's (1981) proposals. He presents a model where determination of three life history features (life span, propagule longevity and establishment requirements) is sufficient to predict the fate of individual wetland species during different water levels. By combining the three life history features, van der Valk recognizes 12 basic life history types. The wetland environment can then be treated as a sieve which permits the establishment of only certain life history types at any given time. To apply van der Valk's model to an actual wetland, one must determine the potential flora of the wetland, and the life history type of each species. The potential flora includes all species growing in the wetland at a given time, plus all additional species represented by buried seeds.

PART II: SHORELINE VEGETATION AND FLORA

Vegetation

Reliable information on the vegetation of Great Lakes wetlands is scattered and scanty. A large number of unpublished marsh management reports and vegetation surveys exist, but they usually suffer from at least one of the following problems:

(1)   They were designed for the study of one or more animal species, and thus only plant species considered important for the animal species are considered.

(2)   They were not carried out by knowledgeable botanists.   Thus, some of the identification of plant species are clearly wrong and many others are questionable. This situation is complicated by the taxonomic difficulties in major plant groups encountered (e.g., Carex, Juncus, Potamogeton).

(3)   They are unpublished or published in non-refereed journals and are, therefore, difficult to locate and evaluate.

(4)   The methodologies used are rarely explained in detail and/or vary significantly from one report to the next. Many are therefore not cited in this survey.

We have recognized (Figure 5) four broad shoreline vegetation types: wet meadow, strand, marsh and aquatic. Each vegetation type has a specific physiognomy, relationship to water levels and life form of dominant species; we will consider the dominant species associated with each vegetation type. Great Lakes wetlands are very complex, with a large number of species capable of dominating local areas.

Wet Meadow

Wet meadows occur wherever slope and substrate conditions are not too steep or rocky.   They are poorly developed on Lake Superior but are a very characteristic feature of the shores of Lakes Huron and Michigan.   Where slopes are very gentle, shoreline wet meadows can cover vast areas.   Fens occur when the wet meadows have extensive calcareous seepage.   They are similar to wet meadows physiognomically and floristically although often more species rich.   Wet meadows and fens support more species than other wetland communities and contain more than half of the species recorded from Great Lakes wetlands.   A single shoreline fen much less than 1 ha in area, near Methodist Point, Georgian Bay, Ontario, contained 96 species (Reznicek, unpublished data).   Dominant species in wet meadows may include Calamagrostis canadensis, Carex lanuginosa, C. lasiocarpa, C. sterilis, C. stricta, Cladium mariscoides, Deschampsia cespitosa, Equisetum variegatum, Eleocharis elliptica, Juncus balticus, Potentilla fruticosa, Scirpus acutus, S. americanus, S. cespitosus, Solidago ohioensis, and Spartina pectinata.   Many other species are also capable of dominating local areas.

Strand

The strand results from disturbance and erosion
occurring at or near the waterline. It is often absent
but may occasionally be very broad. It occurs on both
depositional and erosional shorelines. Dominants are
mainly annuals in the genera Bidens, Cyperus, Eleocharis,
Juncus, Panicum, Polygonum, and Xanthium although a number
of other taxa may occur at a lower frequency.

Marsh

In marshes, emergent species are prominent. Some
emergents can occur in water to ca. 1.5 cm deep, although
best development normally occurs in shallower water.
In most areas, the major dominant is Typha spp. Locally,
Decodon verticillatus, Eleocharis smallii, Phragmites
australis, Pontederia cordata, Sagittaria latifolia, Scirpus
acutus, S. fluviatilis, and Sparganium eurycarpum can
dominate more or less extensive areas. Numerous other
species may also form stands and dominate small areas.
In shallow water (less than ca. 15 cm deep), Carex aquatilis,
C. atherodes, Leersia oryzoides, Lythrum salicaria and
Phalaris arundinacea, as well as some other less important
species, may dominate.

Aquatic

The submersed and floating-leaved aquatic plants
are the least well known of all the Great Lakes wetland
species. They occur in shallow water in openings in marshes,
as an occasional "understory" to emergent aquatic vegetation
and in water deeper than the maximum depth tolerated by
emergent species. The maximum depth to which aquatics
occur is uncertain, but it is at least 8 m (Meyer et al.,
1943; Voss, 1972) under ideal conditions. Species capable
of dominating large areas are numerous. Among the most
important are Ceratophyllum demersum, Elodea canadensis,
Heteranthera dubia, Megalodonta beckii, Myriophyllum spp.,
Najas flexilis, Nymphaea odorata, Nuphar variegatum, Potamogeton
spp., Ranunculus aquatilis (s.l.), Utricularia vulgaris,
and Vallisneria americana.

Flora and Phytogeography

Floristic information covering Great Lakes wetlands
is also scanty. Reliable floristic surveys are available
only for a few local areas, and thus the information presented
here is based mainly on our own field experience. The
flora is quite rich, with about 400-450 species of vascular

plants regularly occurring in Great Lakes wetlands. The most important genera (based on 10 or more species represented) are Carex (ca. 50 spp.), Cyperus (ca. 10 spp.), Eleocharis (ca. 12 spp.), Juncus (ca. 15 spp.), Polygonum (ca. 10 spp.), Potamogeton (ca. 22 spp.), and Scirpus (ca. 13 spp.).

As well as taxonomic diversity, the flora of Great Lakes wetlands comprises several floristic elements. Many aquatic and wetland plants are very widespread; a few, such as Ceratophyllum demersum, are essentially cosmopolitan. Thus, many Great Lakes wetland species are distributed throughout the entire region.

The extensive wetlands on Lake Erie are especially rich in southern species found only rarely or not at all on the other Great Lakes. Examples of species capable of forming extensive stands and dominating communities include Hibiscus palustris, Nelumbo lutea and Nuphar advena as well as many much rarer species, including Boltonia asteroides, Hibiscus laevis, Sagittaria montevidensis and Senecio glabellus.

The wetlands of the southern Great Lakes also have a rich wet prairie element in the flora. This element is especially prominant in the St. Clair River delta marshes (Hayes, 1964). Species representing this element are Helianthus spp., Platanthera leucophaea, Pycnanthemum spp., Solidago riddelli, Veronia spp. and Veronicastrum virginicum.

The fens and wet meadows of the northern portions of Lakes Huron, Michigan and Superior have many boreal, subarctic and (on Lake Superior) arctic species (Given and Soper, 1981). Examples include Carex capillaris, Pinguicula vulgaris, Selaginella selaginoides and Scirpus cespitosus.

CONCLUSION

The existing shoreline vegetation of the Great Lakes depends upon regular fluctuation in water levels. Fluctuating water levels not only increase the area of shoreline vegetation, but increase the diversity of vegetation types and plant species. High water periods prevent woody vegetation and terrestrial species from occupying sites close to the water and temporarily change the vegetation from wet meadow to emergent species, or from emergent species to floating-leaved and submersed species. High water period also kill dominant species such as cattails (Typha spp.)

which might otherwise form extensive monocultures. Low water periods allow many mud flat annuals, meadow and emergent marsh species to regenerate from buried seeds. It appears that buried seed reserves on lakeshores have higher densities than marshes and are more shallow. Any stabilization of water levels would likely reduce marsh area, vegetation diversity and plant species diversity.

Priorities for Future Research

(1)   Classification of major vegetation types.

(2)   Establishment of permanent quadrats to monitor changes in species composition with fluctuating water levels.

(3)   Survey of buried seed reserves in different vegetative types of the Great Lakes.

  (a)   comparison of densities with other wetland types
  (b)   comparison of vertical profile with other wetland types
  (c)   changes along gradients: water depth, exposure to waves
  (d)   list of species persisting as buried seeds

(4)   Comparative studies of flooding tolerance for at least the dominants found in wetlands, with particular emphasis on the depth and duration of flooding required to cause death.

(5)   Investigation of the potential interaction between high water levels, woody plants and the landward limits of marsh vegetation.

(6)   Use of 1-4 to describe cyclic changes in vegetation, in order to predict vegetation responses to different water levels. Testing applicability of van der Valk's Gleasonian hypothesis (van der Valk, 1981).

(7)   Use of 1-5 to predict potential changes in area of wetlands (or of specific wetland types) if water level fluctuations are increased or decreased.

(8)   Investigation of effects of seasonal water level fluctuations upon vegetation diversity.

Two additional refinements would be:

(9)   Determine vegetation response to different frequencies and amplitudes of fluctuations, recognizing that these are two independent components of disturbance.

(10)   Consider rate of recovery from disturbance and determine whether different vegetation types require different intensity/frequency of disturbance to be maintained.

ACKNOWLEDGMENTS

We thank A. Payne for assistance in tabulating the seed bank data, and D.F. Brunton, P. Catling and D. Moore for constructive criticism on drafts of this manuscript. This work was supported in part by a Natural Sciences and Engineering Research Council of Canada operating grant.

DISCUSSION

By

Glenn Guntenspergen
Department of Biological Sciences
University of Wisconsin-Milwaukee
Milwaukee, Wisconsin 53201

Keddy and Reznicek have offered a descriptive model of plant community dynamics in Great Lakes wetlands based on studies in other North American wetlands. In their model, high water causes the elimination of plant species and also makes possible the invasion of plant species into other vegetation communities. Subsequent lower water levels expose the substrate and create conditions that promote the germination of the buried seed pool. This initiates a regeneration of the marsh. Continued fluctuations in water level result in cyclic community development--the pulse stability model put forth by Odum (1971).

A large body of natural history observations do suggest that fluctuating water levels in the semiregulated Great Lakes influence the vegetation dynamics of coastal wetlands. But few studies have quantitatively followed water level fluctuations and changes in plant composition and community distribution (however, see McDonald, 1955 and Harris et al., 1981). Changes in the extent and composition of Great Lakes wetland plant communities are more often inferred from aerial photography (Stearns and Keough, 1982; Harris et al., 1977; Herdendorf et al., 1981) or from historical records (McDonald, 1955; Howlett, 1975).

How well does the hydrologic regime of Great Lakes wetlands resemble their inland counterparts? Are models developed for these systems applicable to coastal systems?

Certainly, water levels are an important environmental factor which affects wetland vegetation. And both coastal and inland wetlands are exposed to fluctuating water levels. However, the frequency and amplitude of these fluctuations may differ. Cycles of 5-40 years are suggested for Iowa wetlands (van der Valk and Davis, 1978), and an irregular 10-20 year cycle has been suggested for the Great Lakes (Bedford et al., 1976). One important difference is that prairie pot holes and waterfowl management areas exhibit fairly major fluctuations in water depth and can become dry within a single year. Coastal Great Lakes wetlands are usually distributed along a sloping soil gradient and experience less drastic fluctuations within any one year. In many Lake Michigan wetlands, at least, a change in water level would not mean the elimination of any vegetation type. More likely, a displacement of the communities (individual species) would occur along the coenocline. However, major disturbances may occur where the slope is steeper or less continuous. The maximum range in water depth on Lake Michigan has been 1.82 m (and only 1.67 m over one cycle) (Bedford et al., 1976). These fluctuations seem to hold true among the other Great Lakes with the exception of Lake Superior. Lake Superior has experienced a maximum range of water depths of only 0.76 m throughout the historical period (Bedford et al., 1976).

In the Great Lakes, the usual seasonal pattern is a winter decline in water level. However, higher winter water levels are also possible and can result in the elimination of certain plant species (see Geis, Chapter 2, this volume and McDonald, 1955). Periods of shallow water may also be followed by unexpected water level fluctuations, which may influence the survival of newly established seedlings. McDonald (1955) studied the recovery of a Lake Erie wetland after high winter water levels (1945-1946). By 1950, only 20 percent of the marsh area had been recolonized and that mostly by vegetation reproduction. Certain decreases in water depth may not stimulate the germination of seeds but may enhance the spread of individuals by vegetative growth.

Herdendorf et al. (1981) proposed a model of plant community displacement for a Lake Michigan wetland with a constant topographic gradient. In their model, plant communities overlap one another along the water depth gradient. This suggests that the distribution of plant communities expands and contracts with fluctuating lake levels as opposed to a wholesale reassortment of species during a drawdown.

The Great Lakes also experience seiche activity (Mortimer and Fee, 1976). Seiches, coupled with storm activity during certain times of the year, may also be a source of disturbance for plant communities. This could include the rafting of vegetation, loss of substrate or breaching of barriers which help to define the wetland.

Geis (see Chapter 2, this volume) and Herdendorf et al. (1981) point out that both shoreline morphology and hydrologic regime define potential wetland types in the Great Lakes. The vegetation present in these wetlands is also a function of other physical factors including topography and substrate as well as biotic factors such as herbivory and competition. In order to completely understand the dynamics of Great Lakes wetlands, we must also include biotic factors and interactions.

Keddy and Reznicek emphasized the role of disturbance and the importance of the seed bank as a partial explanation for the distribution and abundance of plants in Great Lakes wetlands. Low water levels may initiate the germination of a buried seed pool, but the eventual establishment of a plant community depends on the response of the seedlings to subsequent environmental and biotic factors as well as the interaction among adult plants. Competitive interactions among species and specialization on specific resources cannot be dismissed. Experimental work with Typha (Grace and Wetzel, 1981) and studies of the impact of nutrients on wetland communities (Guntenspergen, 1984) suggest that competition can influence the composition of wetland communities. Unfortunately, the biotic processes involved in the degeneration and regeneration of Great Lakes wetlands has not been adequately studied.

The plant communitites of Great Lakes wetlands have been largely neglected, and our knowledge is fragmentory at best. Only sporadic progress has been made since an early attempt at predicting the impact of water level management on Great Lakes wetlands (Bedford et al., 1976). As a prelude to informed management of coastal wetlands and an understanding of their role in supporting wildlife and fishery resources, we need to better understand the processes that control the structure and function of these systems.

LITERATURE CITED

Barclay, A.M. and R.M.M. Crawford. 1982. Plant growth and survival under strict anaerobiosis. J. Exp. Bot. 33:541-549.

Bedford, B., R. Emanuel, J. Erickson, S. Rettig, R. Richards, S. Skavvoneck, M. Vepraskas, R. Walters, and D. Willard. 1976. An analysis of the International Great Lakes Levels Board Report on regulation of Great Lakes water levels: Wetlands, fisheries, and water quality. RF Monograph 76-04. IES Working Paper 30. Wisconsin Dept. of Administration, State Planning Office, Madison, WI. 92 pp.

Bernatowicz, S. and J. Zachwieja. 1966. Types of littoral found in the lakes of the Masurian and Suwalki lakelands. Ekologia Polska - Seria A. 14:519-545.

Bristow, J.M., A.A. Crowder, M.R. King, and S. Vanderkloet. 1977. The growth of aquatic macrophytes in the Bay of Quinte prior to phosphate removal by tertiary sewage treatment (1975-1976). Naturaliste Canadien. 104:465-473.

Connell, J.H. 1978. Diversity in tropical rain forests and coral reefs. Science. 199:1302-1310.

Davidson-Arnott, R.G.D. and W.H. Pollard. 1980. Wave climate and potential longshore sediment transport patterns, Nottawasaga Bay, Ontario. J. Gr. Lakes Res. 6:45-67.

Dore, W.G. and J.M. Gillett. 1955. Botanical survey of the St. Lawrence seaway area in Ontario. Botany and Plant Pathology Division, Canada Department of Agriculture, Ottawa.

Fahselt, D. and M.A. Maun. 1980. A quantitative study of shoreline marsh communities along Lake Huron in Ontario. Can. J. Plant Sci. 60:669-678.

Farney, R.A. and T.A. Bookhout. 1982. Vegetation changes in a Lake Erie marsh (Winous Point, Ottawa County, Ohio) during high water years. Ohio J. Sci. 82:103-107.

Given, D.R. and J.H. Soper. 1981. The arctic-alpine element of the vascular flora at Lake Superior. National Museums of Canada, Publications in Botany No. 10.

Grace, J.B. and R.G. Wetzel. 1981. Habitat partitioning and competitive displacement in cattails (Typha): Experimental field studies. American Naturalist. 118:463-474.

Grime, J.P. 1979. Plant Strategies and Vegetation Processes. John Wiley and Sons, Chichester, UK.

Grubb, P.J. 1977. The maintenance of species richness in plant communities: The importance of the regeneration niche. Biological Reviews of the Cambridge Philosophical Society. 52:107-145.

Guntenspergen, G.R. 1984. The influence of nutrients on the organization of wetland plant communities. Ph.D. Thesis. University of Wisconsin-Milwaukee, Milwaukee, WI. 191 pp.

Haag, R.W. 1983. Emergence of seedlings of aquatic macrophytes from lake sediments. Can. J. Bot. 61:148-156.

Harris, H.J., T.R. Bosley, and F.D. Roznik. 1977. Green Bay's coastal wetlands: A picture of dynamic change. In: C.B. DeWitt and E. Soloway (eds.), Wetlands, Ecology, Values, and Impacts. Proc. of the Waubesa Conference on Wetlands. Institute for Environmental Studies, University of Wisconsin-Madison, Madison, WI. pp. 337-358.

Harris, H.J., G. Fewless, M. Milligan, and W. Johnson. 1981. Recovery processes and habitat quality in a freshwater coastal marsh following a natural disturbance. In: B. Richardson (ed.), Proc. Midwest Conf. on Wetland Values and Management. Minnesota Water Planning Board, St. Paul, MN. pp. 363-379.

Harris, S.W. and W.H. Marshall. 1963. Ecology of water-level munipulations on a northern marsh. Ecology. 44:331-343.

Hayes, B.N. 1964. An ecological study of a wet prairie on Harsen's Island, Michigan. Michigan Botanist. 3:71-82.

Herdendorf, C.E., S.M. Hartley, and M.D. Barnes (eds.). 1981. Fish and Wildlife Resources of the Great Lakes Coastal Wetlands Within the United States. Volume 1: Overview. Biological Services Program, U.S. Fish and Wildlife Service. FSW/OBS-81/02-V1. United States Dept. of Interior, Washington, DC. 469 pp.

Howlett, G.F. 1975. The rooted vegetation of west Green Bay with reference to environmental change. M.S. Thesis. Syracuse University, College of Environmental Science and Forestry, Syracuse, NY.

Huston, M. 1979. A general hypothesis of species diversity. Amer. Nat. 113:81-101.

Hutchinson, G.E.   1975.   A Treatise on Limnology.   Vol. 3. Limnological Botany. John Wiley and Sons, NY.

Jaworski, E., C.N. Raphael, P.J. Mansfield, and B.B. Williamson. 1979. Impact of Great Lakes Water Levels on Coastal Wetlands. Dept. of Geography-Geology, Eastern Michigan University, Ypsilanti, MI.

Jeglum, J.K., A.N. Boissonneau, and V.F. Haavisto. 1974. Toward a wetland classification for Ontario. Information Report O-X-215, Canadian Forestry Service, Department of Environment.

Kadlec, J.A.   1962.   Effects of a drawdown on a waterfowl impoundment. Ecology. 43:267-281.

Keddy, P.A.   1982.   Quantifying within-lake gradients of wave energy: Interrelationships of wave energy, substrate particle size and shoreline plants in Axe Lake, Ontario. Aquatic Bot. 14:41-58.

Keddy, P.A.   1983.   Shoreline vegetation in Axe Lake, Ontario: Effects of exposure on zonation patterns. Ecology. 64:331-344.

Keddy, P.A.   1984.   Quantifying a within-lake gradient of wave energy in Gillfillan Lake, Nova Scotia. Can. J. Bot. 62:301-309.

Keddy, P.A. and A.A. Reznicek. 1982.   The role of seed banks in the persistence of Ontario's coastal plain flora. Am. J. Bot. 69:13-22.

Leck, M.A. and K.J. Graveline.   1979.   The seed bank of a freshwater tidal marsh. Am. J. Bot. 66:1006-1015.

Lyon, J.G. and R.D. Drobney.   1984.   Lake level effects as measured from aerial photos. J. Surveying Eng. 110:103-111.

McDonald, M.E.   1955.   Cause and effects of a die-off of emergent vegetation. J. Wildl. Mange. 19:24-35.

McManmon, M. and R.M.M. Crawford.   1971.   A metabolic theory of flooding tolerance: The significance of enzyme distribution and behavior. New Phytol. 70:299-306.

Meyer, B.F., F.H. Bell, L.C. Thompson, and E.I. Clay. 1943. Effect of depth of immersion on apparent photosynthesis in submersed vascular aquatics. Ecology. 24:393-399.

Moore, J.M. and R.W. Wein. (in press). Soil seed reserves by depth in a Typha latifolia dominated freshwater marsh. Aquatic Bot.

Mortimer, C.H. and E.J. Fee. 1976. Free surface oscillations and tides of Lake Michigan and Superior. Phil. Trans. R. Soc. Lond. A. 281:1-61.

Nicholson, A. and P.A. Keddy. 1983. The depth profile of a shoreline seed bank in Matchedash Lake, Ontario. Can. J. Bot. 61:3293-3296.

Odum, E.P. 1971. Fundamentals of Ecology. 3rd Edition. W.B. Saunders and Company, Philadelphia, PA. 547 pp.

Ponnamperuma, F.N. 1972. The chemistry of submerged soils. Advances in Agron. 24:29-96.

Salisbury, E. 1970. The pioneer vegetation of exposed muds and its biological features. Philosophical Transactions of the Royal Society of London, Series B. 259:207-255.

Sculthorpe, C.D. 1967. The Biology of Aquatic Vascular Plants. Edward Arnold Ltd., London.

Smith, L.M. and J.A. Kadlec. 1983. Seed banks and their role during the drawdown of a North American marsh. J. Appl. Ecol. 20:673-684.

Spence, D.H.N. 1982. The zonation of plants in freshwater lakes. Adv. Ecol. Res. 12:37-125.

Stearns, F. and J. Keough. 1982. Pattern and function in the Mink River watershed with management alternatives. Contract No. POPADA-01417. Wisconsin Coastal Zone Management Program. 101 pp.

Stuckey, R.L. 1975. A floristic analysis of the vascular plants of a marsh at Perry's Victory Monument, Lake Erie. Mich. Bot. 14:144-166.

Thoreau, H.D. 1854. Republished in 1965 as Walden and Civil Disobedience, Airmont Pub., NY.

van der Valk, A.G. 1981. Succession in wetlands: A Gleasonian approach. Ecology. 62:688-696.

van der Valk, A.G. and C.B. Davis. 1976. The seed banks of prairie glacial marshes. Can. J. Bot. 54:1832-1838.

van der Valk, A.G. and C.B. Davis. 1978. The role of seed banks in the vegetation dynamics of prairie glacial marshes. Ecology. 59:322-335.

van der Valk, A.G. and C.B. Davis. 1979. A reconstruction of the recent vegetational history of a prairie marsh, Eagle Lake, Iowa, from its seed bank. Aquatic Bot. 6:29-51.

Voss, E.G. 1972. Michigan Flora. A Guide to the Identification and Occurrence of the Native and Naturalized Seed-plants of the State, Part 1. Gymnosperms and Monocots. Cranbrook Inst. Sci. Bull. 55, Cranbrook Inst. Sci., Bloomfield Hills, and University of Michigan Herbarium.

White, P.S. 1979. Pattern, process and natural disturbance in vegetation. Bot. Rev. 45:229-299.

CHAPTER 4

PRELIMINARY OBSERVATIONS ON THE SEICHE-INDUCED
FLUX OF CARBON, NITROGEN AND PHOSPHORUS
IN A GREAT LAKES COASTAL MARSH

Paul E. Sager
University of Wisconsin-Green Bay
Green Bay, Wisconsin 54301-7001

Sumner Richman
Lawrence University
Appleton, Wisconsin 54912

H. J. Harris
University of Wisconsin-Green Bay
Green Bay, Wisconsin 54301-7001

Gary Fewless
University of Wisconsin-Green Bay
Green Bay, Wisconsin 54301-7001

INTRODUCTION

The exchange of inorganic and organic materials
between wetlands and adjacent waters has been studied
extensively in saltwater systems. Investigations on freshwater
marshes have also been made; however, few such systems
lend themselves to flux measurements of the type made
on estuarine salt marshes. Hence, much of what we do
or do not understand about nutrient dynamics in coastal
marshes comes from salt marsh studies where a more extensive
literature has accumulated. Recent reviews suggest that
the long standing paradigm of outwelling of biologically
important substances, dissolved and particulate, from
coastal marshes cannot be supported (Nixon, 1980; Nixon,
1981; Odum et al., 1979). The reader is referred to these
papers (especially Nixon, 1980) for background on detailed
studies leading to the present understanding that coastal

marshes appear to be nutrient transformers, exporting small amounts of carbon, nitrogen and phosphorus which may not be significant for the receiving waters and associated biota.

A variety of physical factors including the geomorphology of the marsh drainage, the areas of marsh and adjacent coastal waters and the magnitude of the water flux appear to be important determinants of whether specific wetlands show significant export or import of dissolved or particulate substances (Odum et al., 1979). Hence, the variability in physical setting of marshes makes it difficult to compare results of marsh studies and to thereby make useful generalizations. Nixon (1980) attempted to do so in his review and perhaps has come closest to identifying a new paradigm. He observes that coastal marshes appear to annually export both dissolved and particulate organic carbon, dissolved organic nitrogen and dissolved phosphorus. The potential significance of these exports is suggested to be a function of the relative sizes of marsh and coastal water systems; yet in general and based on a variety of study areas, the magnitudes of the exports do not appear to have great biological importance in the adjacent waters.

Two additional studies appear to have particular relevance at this time in the progression of our understanding of the dynamics of coastal marshes. Both studies examine the nature and apparent origin of detritus exported from coastal salt marshes. Harris et al. (1980) noted that the suspended particulates collected from ebb tide waters in a Florida coastal marsh are composed not of vascular plant fragments but of amorphous aggregates, derived primarily from organic films produced by benthic microflora. The second study, by Haines (1977), examined the carbon isotope composition of seston in Georgia estuaries and concluded that this material is derived not from the vascular plants of the salt marsh but from algal production in the estuaries. This implies a minor contribution of marsh particulates to the organic carbon of coastal waters and suggests that the important outflux may be in dissolved form. The studies imply a diminished importance from previous perceptions of the vascular plants as sources of the exported particulate carbon from coastal marshes. They also suggest that in situ decomposition of vascular plant material may be more complete than we believed and that benthic microflora play an important role in carbon cycling in the marsh.

The coastal marshes in Green Bay offer an opportunity to test in a freshwater system the paradigm arising from salt marsh studies and to determine the contribution, if any, these coastal marshes make to the lacustrine ecosystem.

This paper is a preliminary report on a study of a segment of Peter's Marsh on lower Green Bay. The study was initiated in June, 1983. The object of the investigation is to assess the flux of carbon, nitrogen and phosphorus between the marsh and the waters of Green Bay and determine the potential value of exported particulates for filter-feeding zooplankton species of the adjacent open waters.

The study was designed to take advantage of periodic water level fluctuations associated with a standing wave or surface seiche in the bay. Heaps et al. (1982) noted the mean period for the seiche to be 10.8 hrs at the southern end of the bay and that amplitude was observed to range as high as 25-30 cm. Hence a regular water level change, comparable in effect to tidal changes in salt marshes, provides a mechanism to drive an exchange of substances between marsh and bay.

A part of Peter's Marsh, the East Channel Drainage, was selected for the study after observing the extent of flooding and the pattern of water movement through the main channel (Figure 1) during several seiche periods.

METHODS

A recording water level meter was installed at the head end of the drainage channel to monitor seiche activity. Water samples were obtained with a battery powered, automatic sampler located about 10 m from the water level recorder. Samples were taken hourly over 24 hr periods from a depth of 0.5 m in a water column of approximately 1.5 m depth. Water samples taken during periods of very low seiche amplitude or during irregular water level fluctuations were not kept for analysis of carbon, nitrogen and phosphorus compounds.

Analyses of nitrogen and phosphorus compounds were done on a Technicon auto-analyzer following Standard Methods (1975). Carbon analyses were carried out on a total organic carbon analyzer (Oceanography International Corp.).

RESULTS AND DISCUSSION

During the 1984 sampling, the mean amplitude and mean period for 269 seiche events were 19.3 cm and 9.9 hrs, respectively. The seiche was observed to vary in both period and amplitude, the latter ranging as high as 48.5 cm.

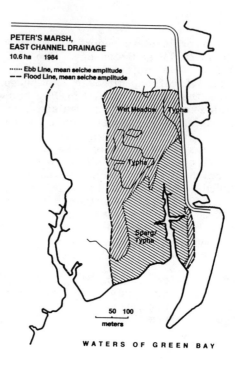

Figure 1. The East Channel Drainage of Peter's Marsh
flood and ebb lines demarcate the areal water flux of
an average seiche.

An example of a data set produced from samples taken
in July, 1983, and matched to the time of the seiche event
is given in Figure 2. Similar data displays are produced
from samples taken during all acceptable seiche events
throughout the study.

A first attempt at data analysis involved reduction
of the full data set to a series of mean values for each
parameter at those sampling times corresponding to the
ebb and flood points of the seiche. This averaging procedure
was done for all dates which had agreement in the time
samples were taken and the time of maximum (flood) and
minimum (ebb) water level for a given seiche. Results
of this data reduction are given in Figures 3 and 4 for
each of four months of sampling.

For the nitrogen and phosphorus data, short-term
effects of marsh ecosystem metabolism are evident in the
differences of ebb and flood values. The e/f ratio may
be used as a measure of net retention (<1) or net release

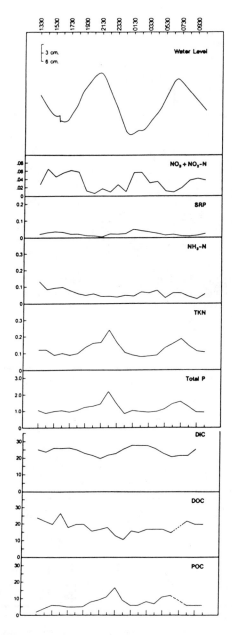

Figure 2. Fluctuations in water level and concentrations of carbon (ug/l) and nitrogen and phosphorus (mg/l) compounds during a seiche event on 26–27 July 1983. Samples taken at the head of East Channel, Peter's Marsh.

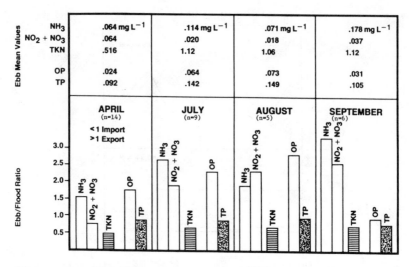

Figure 3. Monthly mean concentrations of nitrogen and phosphorus compounds from samples taken at the head of East Channel, Peter's Marsh, at ebb points of seiche events. Ebb/flood ratios are based on "Ebb" means and similarly derived "Flood" means.

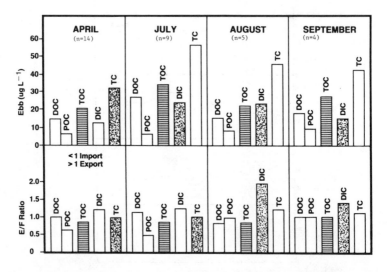

Figure 4. Monthly mean concentrations of carbon compounds from samples taken at the head of East Channel, Peter's Marsh, at ebb points of seich events. Ebb/flood ratios are based on "Ebb" means and similarly derived "Flood" means.

($>$1) of a substance. Ebb values of ammonia-nitrogen were
consistently greater than flood values indicating net
release from the marsh in all four months. The same is
true for nitrite-nitrate-nitrogen except in April when
the e/f ratio indicates net retention. The ratio for
total nitrogen (TKN) suggests a net retention of particulate
(organic) nitrogen forms in the marsh. The subsequent
export or net release of soluble inorganic nitrogen, mostly
as ammonia, but also including nitrite-nitrate forms,
suggests the marsh acts as a nitrogen transformer, partly
in agreement with Nixon's (1980) view of salt marshes;
namely importing dissolved, oxidized forms of nitrogen
and exporting dissolved and particulate nitrogen in reduced
forms.

The phosphorus flux pattern indicated in the e/f
ratios (Figure 3) seems to be in general agreement with
that presented by Nixon. The marsh again appears to behave
as a nutrient transformer, acting as a sink for total
phosphorus and showing a net export of dissolved inorganic
phosphate.

The short term, seiche-induced flux of carbon compounds,
presented in Figure 4 suggests the marsh ecosystem is
also a transformer of carbon. A comparison of ebb and
flood values of particulate organic carbon (POC) indicates
a net retention for most of the sampling period. The
e/f ratio for POC in September represents an interesting
departure from early summer values; however, a t-test
for the ebb-flood difference in September is not significant.
Overall, the marsh appears to import POC and release dissolved
inorganic carbon (DIC). The net, seiche-induced flux
of total carbon (TC) based on the grand mean e/f value
for all samples appears to be from marsh to bay. The
difference between ebb and flood grand means of TC is
not large but is judged to be meaningful in light of the
high net retention of POC and the high net export of DIC
throughout the four months. Presumably, carbon dioxide
is a major component of the DIC.

Seasonal changes in seiche flood values for the
above parameters reflect the rich nutrient-algae conditions
in the bay waters which show dramatic fluctuations typical
of hypereutrophic systems (Barica, 1981). Seasonal changes
in corresponding ebb values are considered to reflect
variations in short-term processes of nutrient transformations
and metabolism in the marsh. In general, based on the
e/f ratios, the highest intensity of carbon, nitrogen
and phosphorus transformations in the marsh seems to be
in late summer when DIC release is greatest, DOC and POC
retention is high, ammonia-nitrogen and orthophosphate

exports are high and net retentions of TKN and TP are high.

The significance of these fluxes in relation to the open water ecosystem of the bay remains to be assessed. The hypereutrophic character of lower Green Bay will likely, however, be a dominant term in that assessment. Additional considerations include groundwater and surface water runoff contributions, the long-term variation in water levels observed over periods of years in the Great Lakes and the bed-load transport of particulates between marsh and bay.

DISCUSSION

Thomas M. Burton
Departments of Zoology and
Fisheries and Wildlife
Michigan State University
East Lansing, Michigan 48824

The data presented in this paper are suggestive of some mechanisms for retention and processing of nutrients by open Great Lakes marshes. In general, particulate components are retained by the marsh, and dissolved components are released based on ebb/flood ratios for seiche-induced water movements. The particulate load of Green Bay water appears to be dropped in the marsh as flow rates decrease due to drag of vegetation and sediments in the marsh. These sediment-derived nutrients then appear to be processed by the marsh (microbial and plant mediated processes perhaps) and released back to Green Bay as dissolved carbon, nitrogen and phosphorus. While this mechanism makes some sense, it cannot be proven with the data presented.

In my opinion, there are severe problems with the methods used in this study and several alternate possibilities exist to explain the observations. For example, groundwater seepage into the marsh from adjacent land could be the origin of the dissolved components. In fact, the high levels of dissolved inorganic carbon are very suggestive of high levels of bicarbonates and aggressive carbon dioxide entering from groundwater. The ammonium and nitrate/nitrite-nitrogen could be derived from the same source.

The placement of a single sampling station at the head of the channel also causes some problems. It is possible that the particulate components apparently retained in the marsh are simply derived from channel scour during flood inputs and settle out locally on the ebb portion of the seiche. Would a station out in the marsh proper (out of the channel) show the same pattern? Likewise, is it possible that particulates are transported at higher depths in the water column on the flood cycle and simply settle out to lower depths on the ebb cycle? Would the same pattern be observed at a lower depth in the water column near the bottom of the channel or for samples integrated over the entire water column? Also, do particulates wash up the channel (a concentrating effect) during the flood cycle and simply re-enter the channel laterally as water exits the marsh?

To answer some of these questions, a much more extensive sampling regime would be required. For example, a series of stations spaced from seepage areas to the channel and adjacent to and along the channel would give one a better idea of the flux of materials into and out of the marsh. If chloride levels differ in marsh and groundwater, ratios of chloride inputs and outputs would be instructive or perhaps other "tracers" of water origin could be found.

I congratulate the authors on their suggestive preliminary findings and hope that more extensive quantitative data will follow. This paper certainly demonstrates the problems inherent in obtaining mass balance nutrient data on marshes open to seiche-induced water movements. Nevertheless, such mass balance data are critical to our understanding of these marshes and their influence on the Great Lakes.

LITERATURE CITED

American Public Health Association. 1975. Standard Methods for the Examination of Water and Wastewater. 14th Ed. Washington, DC, 1193 pp.

Barica, J. 1981. Hypereutrophy - the ultimate stage of eutrophication. WHO Water Quality Bulletin 6(4):95-98, 155-156.

Haines, E.B. 1977. The origins of detritus in Georgia salt marsh estuaries. Oikos. 29:254-260.

Harris, R., B. Ribelin, and C. Dreyer. 1980. Sources and variability of suspended particulates and organic carbon in a salt marsh estuary. In: P. Hamilton and K. MacDonald (eds.), Estuarine and Wetlands Processes, Plenum Press, New York, NY.

Heaps, N.S., C.H. Mortimer, and E.J. Fee. 1982. Numerical models and observations of water motion in Green Bay, Lake Michigan. Phil. Trans. R. Soc. Lond. A 306, pp. 371-398.

Nixon, S.W. 1980. Between coastal marshes and coastal waters - A review of twenty years of speculation and research on the role of salt marshes in estuarine productivity and water chemistry. In: P. Hamilton and K. MacDonald (eds.), Estuarine and Wetlands Processes, Plenum Press, New York, NY.

Nixon, S.W. 1981. Remineralization and nutrient cycling in coastal marine ecosystems. In: Estuaries and Nutrients, B. Neilson and L. Cronin (eds.), Humana Press, Clifton, NJ.

Odum, W.E., J. Fisher, and J. Pickral. 1979. Factors controlling the flux of particulate organic carbon from estuarine wetlands. In: Ecological Processes in Coastal and Marine Systems, R.J. Livingston (ed.), Plenum Press, New York, NY.

CHAPTER 5

NUTRIENT CYCLING BY WETLANDS
AND POSSIBLE EFFECTS OF WATER LEVELS

Darrell L. King
Department of Fisheries and Wildlife
Michigan State University
East Lansing, Michigan   48824

INTRODUCTION

One of the biggest problems in discussing wetlands and particularly the nutrient cylces in wetlands is the development of a common currency of communication. According to the classification of Cowardin et al. (1979), wetlands extend from land which is flooded sometime each year to areas covered with two meters of water during the low water period. This definition includes all of the area between what reasonable people could agree is a red ash or spruce forest to systems that reasonable people could agree are ponds. The tremendous variability across the continuum from forest to pond makes impossible any generalization of nutrient dynamics in wetlands. Thus, the first requirement in any discussion of wetland nutrient dynamics is to define clearly the type of wetland being considered. But even this approach does not allow much generalization.

In his extensive review and analysis of 20 years of salt marsh research, Nixon (1980) laid to rest many shibboleths, particularly those dealing with outwelling, but was unable to arrive at much generalization over the relatively narrow range of conditions represented by salt marshes. Overall, it appears that salt marshes are generally sinks for heavy metals, sinks for phosphorus with some variable remobilization and processors and transformers of nitrogen. Nixon concludes with a quotation from Bronowski (1965) calling for careful attention to the "minute particulars" and suggests that more attention to the "minute particulars" will allow more credibility and less comfort when dealing with wetland nutrient dynamics.

The lack of comfort is a function of the great spatial and temporal variability of nutrient dynamics in wetlands, which in turn, is a function of the interacting variabilities of a great many "minute particulars." Chief among these are those factors which interact to control the thermodynamic shifts in the chemical species of import to nutrient cycling.

THERMODYNAMIC CONSIDERATIONS

Some years ago, Baas Becking et al. (1960) assembled essentially all of the values of pH and Eh from the then existing literature and suggested limits to natural environments as shown in Figure 1. Included in Figure 1 as reference points are equilibrium thermodynamic stability lines for nitrate and nitrite, nitrite and ammonia, and sulfate and hydrogen sulfide as well as the general range of conditions for a variety of aquatic systems.

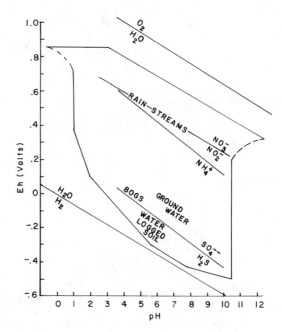

Figure 1. Oxidation-reduction potential (Eh) and pH limits to natural environments (redrawn from Baas Becking et al., 1960). Approximate position of various aquatic environments and thermodynamic equilibrium lines for nitrate:nitrite, nitrite:ammonia and sulfate:hydrogen sulfide pairs.

Figure 1 is a diagram of a primary underlying reason for a great deal of the variability in wetland nutrient dynamics. A July mid-afternoon stroll through a Great Lakes backwater wetland covered with a foot of water reveals the paradox of oxygen bubbles on the leaves of submerged plants and the unmistakable odor of hydrogen sulfide released from the disturbed bottom. The surface waters are characterized by a highly oxidizing environment while the presence of hydrogen sulfide speaks of a much reduced Eh in the bottom sediments (Figure 1). Thus, much of the vertical variability on surface earth noted by Baas Becking et al. (1960) is encompassed in a foot or less of a Great Lakes wetland. By late August the same wetland may be dry with oxygen penetrating the bottom soils. As such, wetlands are not only transitions between the land and the water but also transitions between oxidizing and reducing environments. In a thermodynamic sense wetlands are characterized as highly variable spatial and temporal transitional environments. Much of the variability in nutrient flux noted in wetlands of various types can be traced to this highly variable thermodynamic character.

Both pH and Eh of wetlands are extremely variable functions of biotic activity. Photosynthetic oxygen production in the surface water maintains an elevated Eh while photosynthetic extraction of carbon dioxide from the carbonate-bicarbonate alkalinity raises pH, with the amount of pH increase being a function of nutrient availability, alkalinity and detention time of the water. Bacterial use of the resulting plant tissues and the concomitant release of carbon dioxide reduces both pH and Eh with the rate and degree of change being a function of the rate of water throughput.

HYDROLOGY

Materials are brought to and removed from wetlands largely as a function of water movement, and the pattern of water movement is the primary determinant of the direction and thrust of nutrient flux in these complex systems. Under north temperate climates, seasonal hydrologic variability is characterized by maximum water discharge to wetlands associated with snowmelt and spring rains. Nutrient discharge from the land tends to be correlated with water discharge (Brehmer, 1958; Kevern, 1961; Vannote, 1961), and in most wetlands the major portion of the annual nutrient load enters in the spring. Elevated spring water flows coupled with reduced structural complexity of wetlands in the spring cause much of this nutrient load, together with nutrients stored the previous summer, to be moved rapidly through the wetland.

The general pattern of reduced water discharge during the summer results in increased detention of water in wetlands during this period. However, water seldom runs through wetlands in a uniform manner, but is channeled such that the mean detention time of the water varies greatly throughout the wetland. In those areas where the detention time is increased, the water temperature rises and pH increases because of changes in the carbonate-bicarbonate equilibrium caused by both the warming of the water and increased photosynthetic extraction of carbon dioxide from the alkalinity. In most wetlands both temperature and pH exhibit significant spatial variability as a function of variable water throughput rate compounded by differential plant growth during the summer months.

In those areas where water movement is slowed, biotic utilization of accumulating organics allows establishment of increasingly reducing environments in the wetland sediments. But, here too, the rate and extent of formation of reducing conditions is a function of water movement over and through the bottom materials.

Thus, changes in pH, Eh and temperature, all of which play prime roles in nutrient flux, are dictated to a large extent by the pattern of water movement through the particular wetland.

SEDIMENTS

The general feeling is that sediments accumulate in wetlands; for if they did not, the State of Louisiana would be much smaller than it is. However, for any particular wetland, sediment deposition and retention will be a function of hydrology, wetland structure and sediment type. Generally, sediment accrual will increase with increased water detention time.

Sediments, and particularly inorganic sediments, play significant roles in the ability of wetlands to retain phosphorus and heavy metals. The equilibrium adsorption capacity of sediments for such materials varies as an interactive function of both sediment type and chemical characteristics of the water. An example of these interactions is seen in the equilibrium phosphorus adsorption capacity of three different clays shown by Edzwald (1977) to vary as a function of clay type and water pH in the manner shown in Figure 2. Edzwald concluded that the type and amount of the free metal content of the clay played a major role in determining its phosphorus adsorption potential.

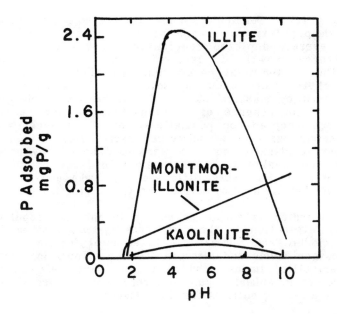

Figure 2. Phosphorus adsorption on illite, montmorillonite and kaolinite as a function of water pH (redrawn from Edzwald, 1977).

Clearly, the ability of wetland sediments to bind materials will be a function of the type of sediment, the water concentration of the materials in question and factors such as pH of the water determined by yet other interactions. The retention of the nutrient-loaded sediments in the wetland is largely a function of hydrology.

BIOTIC INTERACTIONS

Nutrients cycle in wetlands as elsewhere as a function of biotic activity limited by interacting physical, chemical and biological factors. On the broad scale, macrophyte and algal production in wetlands varies as a function of light, temperature and nutrient availability while the remainder of the community is limited by the production and introduction of organic carbon. Within these broad limits, however, biotic interaction can alter the entire ecological structure of shallow water systems. As was noted earlier, biotic activity has a marked effect on the thermodynamic character of wetlands.

Another example of such alteration is given by Spencer and King (1984) who showed marked changes in two meter

deep, nutrient enriched ponds associated with changes
in fish populations. Large numbers of zooplanktivorous
fish severely depleted zooplankton populations allowing
unfettered growth of planktonic algae. The heavy algal
growths limited light penetration such that by midsummer
macrophytes were extremely scarce and those ponds were
dominated by massive blooms of nitrogen-fixing bluegreen
algae. The absence of zooplanktivorous fish resulted
in large zooplankton populations, few algae, clear water
and massive growths of submerged macrophytes. The ponds
without zooplanktivorous fish were used heavily by nesting
ducks and muskrats, while those with large populations
of the small fish were little used by ducks and muskrats
but were heavily used by herons, egrets and kingfishers.

A factor as simple as change in the fish population
caused alteration in these wetlands systems ranging from
type of bird and mammal use to chemical factors such as
pH, oxygen levels and nitrogen fixation. Thus, in addition
to variability imposed by plant and bacterial activity,
differences in animal populations represent another variable
feature affecting nutrient flux in wetlands.

NUTRIENT CYCLES

To this point it has been suggested that nutrient
cycles in wetlands are controlled largely by chemical
thermodynamics and mediated by biotic activity relative
to the inputs of water and material. Seasonal variations
in inputs of both water and materials in north temperate
wetlands coupled with seasonal variation in light and
temperature ensure marked seasonal variation in the nutrient
dynamics of wetlands.

The accelerated production of wetland vegetation
during the spring and summer captures nutrients which
are released as the accumulated vegetation is utilized
during the fall and winter. The typical high water discharge
through the wetland the following spring washes out much
of the released nutrient. Thus, for many wetlands, samples
taken during the summer would indicate that the wetland
is a sink for many nutrients while samples collected during
the early spring would suggest that the wetland is a nutrient
source.

This suggests that, depending on where and when
data are collected from wetlands, information can be accumulated
which indicates that any given wetland is either a sink
or a source for almost any nutrient. But, even in a single
season there will be a large variability in nutrient flux
in most wetlands. For example, during the summer there

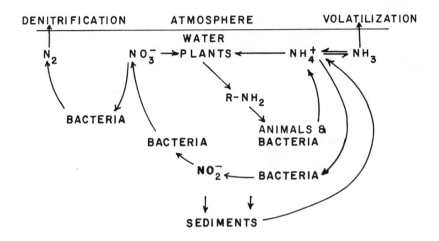

Figure 3.   Nitrogen cycle in aquatic systems.

is tremendous spatial variation in nutrient dynamics within wetlands governed largely by interactions between plant photosynthesis and hydrology. The reasons for such variation can be illustrated with the effects on the nitrogen cycle caused by interactions between plant activity and the hydrology.

Photosynthesis by the mix of submerged, emergent, planktonic and periphytic plants increases both the dissolved and particulate organic content of the wetland. Increased supply of energy-rich organics allows accelerated respiratory activity in the warming water increasing the transformation rate of nitrogen from one chemical species to another (Figure 3).

Inspection of Figure 3 shows two avenues of nitrogen loss to the atmosphere, both of which tend to be accelerated by increased water detention time. Increased water detention time increases the probability of establishing sufficiently reducing conditions to allow denitrification and the loss of nitrogen to the atmoshpere. Decreased water flow rate through a wetland also is accompanied by increased photosynthetic carbon extraction from the alkalinity and the concomitant rise in pH. This pH rise can lead to rapid losses of ammonia to the atmosphere.

Bowden (1984) recently estimated the annual net ammonium production rate of a freshwater tidal marsh in Massachusetts to be about 1.7 mol $N/m^2/yr$ (23.8 g $N/m^2/yr$)

in the top 10 cm of sediment. King (1979), reporting on a series of two meter deep ponds in Michigan dominated by submerged macrophytes and charged with a good quality secondary effluent, noted a 97 percent nitrogen removal in a detention time of 120 days. Nitrogen was lost at the exponential rate of 3 percent per day of detention (equation 1).

$$N_t = N_o \, e^{-0.03t} \qquad (1)$$

Where
N_t = nitrogen (total as N) at any time t (mg/l)
N_o = nitrogen (total as N) entering the system (mg/l)
t   = detention time (days)

Massive harvest of the submerged plants accounted for only about 10 percent of the nitrogen loss with much of the remainder being lost as ammonia gas to the air. An input-output mass balance of nitrogen flux in the first of these shallow, macrophyte dominated ponds during May of 1977 gave a loss rate of 16.3 mg N/m^2/hr. During this period, the pH of the pond ranged from 8.65 to 10.30, the temperature from 15 to 25C and total ammonia in excess of atmospheric equilibrium from 3.50 x 10^{-7} to 3.37 x 10^{-4} moles NH_3/l.

Galloway (1980), using these data and his equation for calculating ammonia flux based on ammonia equilibrium, pH, temperature and wind conditions, calculated an average ammonia loss rate of 15.6 mg N/m^2/hr. Thus, it appears that 96 percent of the 16.3 N/m^2/hr lost from the pond during this period was lost as ammonia. The interaction of pH and water temperature to determine the rate of ammonia loss from a two meter deep system subjected to a wind of two m/sec is given in Figure 4. The values in Figure 4 were calculated with Galloway's equation. It is clear from Figure 4 that the rate of loss of ammonia to the atmosphere is an interactive function of pH and temperature, both of which tend to increase with increased detention time in a wetland.

Thus, increased water detention time, particularly during the summer, will tend to accelerate nitrogen loss to the atmosphere through both denitrification and ammonia volatilization. Clearly, interactions between plant activity and hydrology dictate the amount of nitrogen lost in this manner which, from input-output studies, would allow the wetland to be counted as a nitrogen sink.

Phosphorus dynamics in wetlands are equally complex and involve a variety of shifts in chemical equilibria

(Mortimer, 1941-1942), precipitation kinetics (Stumm and Morgan, 1970) and a variety of biological interactions (e.g., Harrison et al., 1972; McRoy et al., 1972). Much of the dynamics of phosphorus can be traced to interactions between plant activity and hydrology and type and amount of sediments added to the wetland.

RESEARCH NEEDS

   Despite the large numbers of studies of nutrient flux in wetlands, it does not appear that we know much about the nutrient dynamics of these systems. Nixon (1980) was unable to arrive at much generalization about the movement of nutrients through salt marshes. Burton (1981),

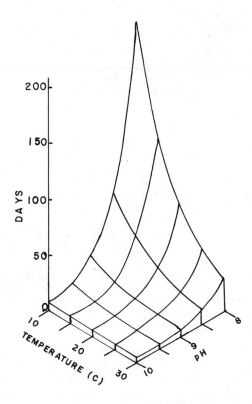

Figure 4. Time (days) required to volatilize one-half of the ammonia in excess of atmospheric equilibrium from the water to the atmosphere as a function of water pH and temperature from a two meter deep system subjected to a surface wind of two m/sec.

in his review of the effects of riverine marshes on water quality, arrived at essentially the same conclusion. Burton noted that most of the studies conducted include evaluation of input and output nutrient concentrations but rarely include any real measures of the input and output of water. Based on the limited data available, Burton concluded that there is a great deal of seasonal variability in the export of phosphorus and nitrogen through riverine marshes but that they are probably sinks for nitrogen and phosphorus on an annual basis. Generally, we are left with the feeling that if sediments are deposited and retained, the wetland probably is a sink for phosphorus. The fate of nitrogen in wetlands is much more questionable.

It seems clear that we will never be able to generalize nutrient cycling across the broad range of wetlands from the red ash forest to the pond. It is equally questionable whether we will ever be able to incorporate all of the "minute particulars" sufficiently to allow a complete description of nutrient flux through a single wetland. In fact, the immense variability from backwater to channelized area within a single wetland makes the collection of statistically reliable nutrient flux data for the entire wetland difficult if not impossible.

In spite of these limits, there is opportunity for research which will improve our knowledge of nutrient cycling in the complex systems represented by wetlands. There is need for research which evaluates the extremes of nutrient flux in wetlands as functions of alterations in the various physical, chemical and biological parameters associated with well-defined hydrologic and vegetative characteristics. There also is need for long-term, carefully conducted studies of total nutrient mass balance in wetlands with near-continuous measurement of both nutrient concentration and water discharge, again, relative to well defined hydrologic and vegetative characteristics. The primary need for both approaches is to recognize those factors which interact to make each wetland unique and to measure nutrient flux as a function of those factors.

Each wetland is unique with a different melange of physical, chemical and biological factors which interact to dictate nutrient flux. Differences in morphology, hydrology, sediment type and nutrient load from wetland to wetland and within single wetlands ensure marked differences in nutrient flux within, among and between wetlands.

It is not possible to include evaluation of all of the "minute particulars" in any research study, but the value of the results of any research on wetland nutrient cycling will be associated directly with the care and attention given to relating nutrient flux to the remainder of the wetland ecosystem studied.   Random measures of nutrient flux in wetlands are of little value; but results from carefully defined studies relating nutrient flux to specific control variables will serve as the base for development of procedures allowing effective management and maintenance of these dynamic, productive systems.

DISCUSSION

Jeffrey M. Kaiser
#408 - 3120 Kirwin Ave.
Mississauga, Ontario

I agree with Professor King's observations that wetlands are:

1) an extremely heterogeneous group of systems across the face of this planet and even over the extent of North America;

2) that they are pulse-fed on temporal cycles which are neither regular in occurrence nor volume, nor do these pulses uniformly impact the recipient wetlands;

3) that each individual wetland is spatially heterogeneous in respect to physical factors such as water regime, substrate type and nutrient chemistry and such biological factors as species and community distribution and relative biological activity (such as turnover rate and productivity); and

4) that between wetlands there is seasonal variation in such critical physical factors as climatology and hydrology and biological factors as production rates and standing crop.

Even within the subset of wetlands on the shores of the North American Great Lakes, the above observations apply.   But having set out this list of problems, the task of discerning common patterns and processes in nutrient cycling is not impossible, though it provides a formidable challenge.   Perhaps we are well advised to consider that "what is past is prologue" and we must now begin to accumulate the many observations and careful measurements which will provide us with a good basis from which we can consider the many implications for wetland use, management and manipulation which nutrient cycles have.

Generally, the Great Lakes lie in a region of similar geology, climatology, physiography and biology, if we recognize a major distinction between the Upper and Lower Lakes and a lesser distinction between north and south shores (Chapman and Putnam, 1972; Brown et al., 1974; Chapman and Thomas, 1968; Herdendorf et al., 1981). Furthermore, nearly all of the wetlands are subjected to the important process of water drawdown as watershed run-off drops from annual peaks in March and April to lows in September (Environ. Canada, n.d.) and lake levels drop from June peaks to January and February lows (Jaworski et al., 1979). Superimposed on this annual cycle are the short-term inundations of regular but unpredictable episodes of rainfall and the occurrences of seiches. The effects of all or any of these events on a given wetland, of course, depend on the type of wetland and its location.

It appears to me that hydrology, the fluxes and storage of water, is the single most important factor in nutrient cycling, determining most elemental inputs, detention time and the bulk of the outputs. The hydrological regime is most subject to modification by the morphometry of a given wetland and, after that, by the biological components of the wetland system.

Of the various morphometric classifications of wetlands, a relatively simple one such as that used by the International Joint Commission is quite adequate for hydrological considerations (International Great Lakes Levels Board, 1973). It recognizes seven types of wetlands from open shoreline to restricted riverine to protected (by barrier beach or dike).

Hydrology can be conceptualized according to a model such as Figure 5, recognizing that certain modifications will be needed when considering a particular morphological type. In many instances the same model can be used to conceptualize the general fluxes and storage of nutrients. For most nutrients, with the probable exception of nitrogen, the bulk of their input occurs in telluric waters. However, because of their proximity to large population and industrial centers, a substantial number of wetlands may receive large atmospheric inputs of combustion by-products such as sulfur and, hence, neither rainfall nor dry deposition can be summarily ignored. Nor can the role of the seiche be discounted if nearby portions of the lake receive industrial or municipal effluents.

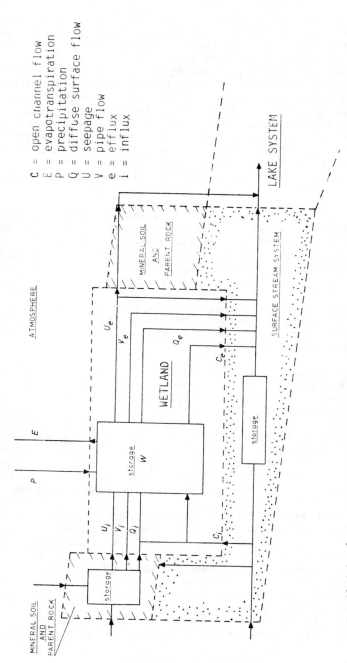

Figure 5. Model of the hydrological relations of a wetland showing the types of fluxes which occur. With some modification the model can also be applied to many nutrient fluxes. (Modified from Crawford, 1983).

We can look at the enormous output of scientific and technical information over the last 20 years and wonder why so little is known about nutrient cycling. The reasons are not obscure. The earliest reliable studies were only made in the 1950s (Hutchinson and Bowen, 1950; Rigler, 1956) and the majority of subsequent studies were performed in limnetic systems where compartmentalization was relatively straightforward (Likens, 1975). The primary focus of the limnological work has been nitrogen (e.g., Brezonik, 1973) or phosphorus (e.g., Rigler, 1973), and it should provide a good jumping off point for studies in wetlands. Undoubtedly, wetlands will have more storage "compartments" and a more complex set of transfer routes than do most lake systems, but that is part of the challenge in studying these systems.

For some of the compartments, a good background has been provided for further research. Ponnamperuma (1972) has provided an excellent treatment of the chemistry of flooded soils; Yoshida (1975) has reviewed microbial metabolism in flooded soils; Crawford (1983) and Yoshida and Tadano (1978) have reviewed the adaptations of and metabolism by higher plants in flooded soils. Rather few studies of the more general aspects of nutrient cycling in wetlands are available but those by Klopatek (1975) and Kitchens et al. (1975) provide some interesting data for comparative purposes.

I would like to conclude with two final points. The majority of nutrient cycling studies have dealt with the elements nitrogen and phosphorus. Whether carbon is considered a primary nutrient or not, it, too, has received considerable attention from limnologists because of its role in aquatic chemistry and secondary production (e.g., Wetzel, 1975, Chap. 10). The reason for the interest in nitrogen and phosphorus is quite appropriate given their frequent roles as limiting nutrients and the accelerating alteration of natural systems due to human additions of these elements. However, we must not lose sight of the fact that boron, calcium, chlorine, cobalt, copper, iodine, iron, magnesium, manganese, molybdenum, potassium, sodium, sulfur, vanadium and zinc are also essential nutrients and that they, too, play a role in controlling the rate of nutrient cycling. Thus, Goldman (1960) found a lake in California where molybdenum apparently limited primary production and another in Alaska (Goldman, 1972) where magnesium was limiting. In a recent study of Clear Lake, California, Wurtsbaugh and Horne (1983) found that iron limited algal productivity. Hopefully, students of wetland nutrients will also consider the behaviour of the "minor" nutrient elements.

Finally, I would like to suggest that though wetlands may and probably do alter the rate and timing of nutrient effluxes from that of elemental influxes, so long as such wetlands serve as sediment and detrital traps, they must be considered sinks. The fraction of total input which accumulates and the rate of accumulation will vary for different elements, but all except the most mobile such as chlorine, sodium and perhaps iodine must eventually enter and become sequestered in wetland sediments.

LITERATURE CITED

Baas Becking, L.G.M., I.R. Kaplan, and D. Moore. 1960. Limits of natural environment in terms of pH and oxidation-reduction potential. J. Geol. 68:243-284.

Bowden, W.B. 1984. A nitrogen-15 isotope dilution study of ammonium production and consumption in a marsh sediment. Limnol. Oceanogr. 29:1004-1015.

Brehmer, M.L. 1958. A study of nutrient accrual, uptake and regeneration as related to primary production in a warm-water stream. Ph.D. Thesis. Michigan State University, East Lansing, MI. 97 pp.

Brezonik, P.L. 1973. Nitrogen sources and cycling in natural waters. U.S.E.P.A. Report 660/3-73-002.

Bronowski, J. 1956. Science and Human Values. Harper and Row. 119 pp.

Brown, D.M., G.A. McKay, and L.J. Chapman. 1974. The climate of southern Ontario. Environment Canada. Atmos. Environ.

Burton, T.M. 1981. The effects of riverine marshes on water quality. In B. Richardson (ed.), Selected Proceedings of the Midwest Conference on Wetland Values and Management. Minn. Water Planning Board, p. 139-151.

Chapman, L.J. and D.F. Putnam. 1972. The physiography of southern Ontario. 2nd ed., Univ. Toronto Press.

Chapman, L.J. and M.K. Thomas. 1968. The climate of northern Ontario. Canada Dept. Transport. Meteorological Branch.

Cowardin, L.M., V. Carter, F.C. Golet, and E.T. LaRoe. 1979. Classification of wetlands and deepwater habitats. USFWS. FWS/OBS-79/31, 103 pp.

Crawford, R.M.M. 1983. Root survival in flooded soils. In A.J.P. Gore (ed.), Mires: Swamp, Bog, Fen and Moor. pp. 257-283.

Edzwald, J.K. 1977. Phosphorus in aquatic systems: The role of the sediments. In I.H. Suffet (ed.), Fate of Pollutants in the Air and Water Environments, Part 1, Vol. 8. John Wiley and Sons, New York, NY. pp. 183-214.

Environment Canada, n.d. What you always wanted to know about Great Lakes levels. Canada Dept. Environment.

Galloway, J.E. 1980. Aquatic nitrogen cycling: The processing of nitrate by Chlorella vulgaris and the possibility of ammonia loss to the atmosphere. Ph.D. Thesis. Michigan State University, East Lansing, MI. 157 pp.

Goldman, C.R. 1960. Molybdenum as a factor limiting primary productivity in Castle Lake, California. Science. 132:1016-1017.

Goldman, C.R. 1972. The role of minor nutrients in limiting the productivity of aquatic ecosystems. In G.E. Likens (ed.), Nutrients and Eutrophication: The Limiting Nutrient Controversy. pp. 21-33. Am. Soc. Limnol. Ocean. Spec. Symp. 1.

Harrison, M.J., R.E. Pacha, and R.Y. Morita. 1972. Solubilization of inorganic phosphates by bacteria isolated from Upper Klamath Lake sediments. Limnol. Oceanogr. 17:50-57.

Herdendorf, C.E., S.M. Hartley, and M.D. Barnes (eds.). 1981. Fish and Wildlife Resources of the Great Lakes Coastal Wetlands within the United States. 5 vols. U.S. Fish and Wildlife Service.

Hutchinson, G.E. and V.T. Bowen. 1950. Limnological studies in Connecticut. IX. A quantitative radiochemical study of the phosphorus cycle in Linsley Pond. Ecology. 31:194-203.

Ingram, H.A.P. 1983. Hydrology. In A.J.P. Gore (ed.), Mires: Swamp, Bog, Fen and Moor. pp. 67-158.

International Great Lakes Levels Board. 1973. Regulations of Great Lakes water levels. Report to the International Joint Commission.

Jaworski, E., C.N. Raphael, P.J. Mansfield, and B.B. Williamson. 1979. Impact of Great Lakes water level fluctuations on coastal wetlands. Mich. Office of Water Resources, Tech. Cont. 14-00001-7163.

Kevern, N.R. 1961. The nutrient composition, dynamics and ecological significance of drift material in the Red Cedar River. M.S. Thesis. Michigan State University, East Lansing, MI. 94 pp.

King, D.L. 1979. The role of ponds in land treatment of wastewater. In H.L. McKim (ed.), International Symposium on Land Treatment of Wastewater, Vol. 2. U.S. Army Cold Regions Research and Engineering Laboratory, Hanover, NH. pp. 191-198.

Klopatek, J.M. 1975. The role of emergent macrophytes in mineral cycling in a freshwater marsh. In F.G. Howell, J.B. Gentry, and M.H. Smith (eds.), Nutrient Cycling in Southeastern Ecosystems. pp. 367-393. U.S. ERDA.

Likens, G.E. 1975. Nutrient flux and cycling in freshwater ecosystems. In F.G. Howell, J.B. Gentry, and M.H. Smith (eds.), Mineral Cycling in Southeastern Ecosystems. pp. 314-348. U.S. ERDA.

McRoy, C.P., R.J. Barsdate, and M. Webert. 1972. Phosphorus cycling in an eelgrass (Zostera marina L.) ecosystem. Limnol. Oceanogr. 17:58-67.

Mortimer, C.H. 1941-1942. The exchange of dissolved substances between mud and water in lakes. J. Ecol. 29:280-329; 30:147-201.

Nixon, S.W. 1980. Between coastal marshes and coastal waters--A review of twenty years of speculation and research on the role of salt marshes in estuarine productivity and water chemistry. In P. Hamilton and K.B. MacDonald (eds.), Esturine and Wetland Processes with Emphasis on Modeling. Plenum Publishing Corporation, New York, NY. pp. 437-525.

Ponnamperuma, F.M. 1972. The chemistry of submerged soils. Advances in Agron. 23:29-96.

Rigler, F.H. 1956. A tracer study of the phosphorus cycle in lakewater. Ecology. 37:550-562.

Rigler, F.H. 1973. A dynamic view of the phosphorus cycle in lakes. In E.J. Griffith, A.M. Beeton, J.M. Spencer, and D.T. Mitchell (eds.), Environmental Phosphorus Handbook. pp. 539-572.

Spencer C.N. and King, D.L. 1984. Role of fish in regulation of plant and animal communities in euthrophic ponds. Can. J. Fish. Aquat. Sci. 42:1851-1854.

Stumm, W. and J.J. Morgan. 1970. Aquatic Chemistry. Wiley-Interscience. New York, NY. 583 pp.

Vannote, R.L. 1961. Chemical and hydrological investigations of the Red Cedar River watershed. M.S. Thesis. Michigan State University, East Lansing, MI. 126 pp.

Wetzel, R.G. 1975. Introduction to Limnology. J. Wiley & Sons, New York, NY.

Wurtsbaugh, W.A. and A.J. Horne. 1983. Iron in eutrophic Clear Lake, California: Its importance for algal nitrogen fixation and growth. Can. J. Fish Aquat. Sci. 40:1419-1429.

Yoshida, S. and T. Tadano. 1978. Adaptation of plants to submerged soils. In G.A. Jung (ed.), Crop Tolerance to Suboptimal Land Conditions. pp. 233-256. Am. Soc. Agron. Spec. Publ. 32.

Yoshida, T. 1975. Microbial metabolism of flooded soils. In E.A. Paul and A.D. McLaren (eds.), Soil Biochemistry, 3. pp. 83-123.

CHAPTER 6

AVIAN WETLAND HABITAT FUNCTIONS
AFFECTED BY WATER LEVEL FLUCTUATIONS

Martin K. McNicholl
Long Point Bird Observatory
Port Rowan, Ontario, Canada N0E 1M0

INTRODUCTION

Wetlands provide feeding habitat for a wide variety of birds year round and seasonal habitats for nesting, moulting, migration stop-over sites, and wintering sites. As effects on birds of fluctuations in water levels on suitability of a particular wetland for feeding will be manifest primarily through effects on the food supply or even less directly through effects on the habitat used by the food organism(s) in question, I shall not address this function further here directly, though food supply is an important factor altering suitability of some sites for other functions. This discussion, then, relates primarily to seasonal habitat functions.

The Great Lakes provide important nesting areas for various marsh birds, including waterfowl (e.g., McCracken, 1981; McCracken et al., 1981; Scharf and Shugart, 1984), some shorebirds (McCracken et al, 1981), and some colonial waterbirds (e.g., Blokpoel, 1977; Courtenay and Blokpoel, 1983; Weseloh et al., 1983; Vermeer and Rankin, 1984). Important habitat is also available on the Great Lakes for migration stop-overs (Bradstreet et al, 1977; McCracken, 1981), and waterfowl and larids are known to winter in places in large numbers (e.g., Goodwin, 1982; Lambert, 1981). I am not aware of any reports of large concentrations of moulting birds on the Great Lakes.

NESTING

Habitat Stability

Nesting habitat for waterbirds can be divided according to stability into three broad categories: stable, unstable, and intermediate (McNicholl, 1975). The most stable sites are those on cliffs or on high rocks, well above water line, and these should be seldom if ever affected directly by fluctuations in water level (McNicholl, 1975). Unstable habitats are those such as sand bars, mud banks, and beaches that change frequently and often rapidly in their relation to surrounding water (McNicholl, 1975), with loss of nests to flooding common (see McNicholl, 1975; Clapp et al., 1983, for many references). Intermediate habitats are those which either fluctuate from one year to the next or over a period of years, or which slowly become unsuitable through a progressive change such as vegetative growth or gradual erosion (McNicholl, 1975; Buckley and Buckley, 1980). The best studied habitat of intermediate stability is that of marshes and shallow prairie lakes. Species nesting in these habitats cannot necessarily nest in the same sites from year to year (McNicholl, 1975; Buckley and Buckley, 1980) and show a wide variety of adaptations to nesting in unpredictable conditions (Weller and Spatcher, 1965; Burger, 1980; Orians, 1980; Weller, 1981). Species nesting in more slowly changing habitat may be more inclined to persist when conditions are less than ideal but will have to change sites eventually (McNicholl, 1975; Buckley and Buckley, 1980; Courtenay and Blokpoel, 1983). Some apparently unstable sites may in fact be reliably available at the time of year most needed, provided that the species using them are adapted to commence nesting quickly. For example, a gravel bar in the Amazon River is under water each year, but so reliably emerges annually that the local residents have a place name designated for it, and terns and skimmers start nesting on it soon after the waters recede (Krannitz, 1984).

Nesting Habitat in the Great Lakes

Stable habitats are abundant on the Great Lakes, especially on Lakes Huron and Superior, where numerous islands are used by colonial waterbirds for nesting, as shown by extensive surveys by H. Blokpoel, D. V. Weseloh, and others. Water levels likely affect these colonies only indirectly, and changes in numbers of breeding birds such as the recent increase in double-crested cormorants (Phalacrocorax auritus) probably result primarily from other factors (Weseloh et al., 1983; Vermeer and Rankin, 1984).

Many of these islands are hard granite rocks with relatively little soil suitable for extensive vegetation. A few, however, are heavily vegetated and may become unsuitable for nesting unless flooding occurs (pers. obs.). Courtenay and Blokpoel (1983) attributed shifts of colony sites of common terns (Sterna hirundo) on the lower Great Lakes over the last century at least partly to encroaching vegetation at many traditional nesting areas. These colonies are affected, then, not by fluctuations in water levels, but by lack of the sorts of fluctuations that would cut back the vegetation to an earlier stage.

Although major portions of the Great Lakes are devoid of marshes, extensive marsh habitat occurs in Wisconsin at Green Bay, in various parts of Michigan, and in Ontario on Lakes Erie and St. Clair. These marshes host a wide variety of bird species that are well adapted to nesting in habitat of fluctuating water (McCracken, 1981; McCracken et al., 1981). In addition to species that have long been known to nest in these marshes, Forster's terns (Sterna forsteri) have recently colonized parts of Ontario and Michigan from which they were not previously known (Peck, 1976; Peck and James, 1983; Scharf and Shugart, 1984), and now nest quite abundantly both at Long Point, Ontario (McCracken, 1981; McCracken et al., 1981) and Saginaw Bay, Michigan (Scharf and Shugart, 1984) in addition to their previously known Lake St. Clair and Wisconsin nestings. Little gulls (Larus minutus) have shown some evidence that they may be at the beginning of a similar colonization process (reviewed in McCracken, 1981).

Rivers along the Great Lakes are not known for unstable sand bar nest sites, but beaches are used for nesting in some areas and are subject to the same threats of flooding as elsewhere (McCracken, 1981).

Avian Adaptations to Different Water Levels

Avian communities have been studied in marshes in relation to water levels and habitat differences in various marsh species (e.g., Sowls, 1955; Weller and Spatcher, 1965; Miller, 1968; Orians, 1980; Hochbaum, 1981; McCracken, 1981; McCracken et al., 1981; Weller, 1981). These studies have shown a wide range of adaptations to water depth and vegetation types (Weller, 1981) in spite of the relative simplicity of marsh ecosystems (Orians, 1980; Weller, 1981). Ducks show a wide degree of habitat preference, ranging from dry land sites to deep water (Sowls, 1955; numerous other studies). Black terns (Chlidonias niger) show a much broader habitat tolerance than Forster's terns,

with Blacks nesting in both shallow and deep waters whereas
Forster's usually require deeper, more open water (Weller
and Spatcher, 1965). Red-winged blackbirds (Agelaius
phoeniceus) and yellow-headed blackbirds (Xanthocephalus
xanthocephalus) exhibit a similar difference in level
of tolerance of different habitats, with the red-wing
much more likely to nest farther from open water (Weller
and Spatcher, 1965; Miller, 1968; Orians, 1980). In recent
years, red-winged blackbirds seem to have adapted even
further to non-marsh conditions (Orians, 1980; McNicholl,
1981) and in 1984 appeared to be nesting on the University
of Toronto campus, far from any evident standing water
(pers. obs.). Species of wider habitat tolerance will
obviously be less affected at the population level by
fluctuations in water level than those with more rigid
requirements.

Marsh nesting species, like beach nesting birds,
are subject to considerable nest loss due to flooding
when waters fluctuate (Bennett, 1938; Sowls, 1955; Lynch,
1964; McNicholl, 1979, 1982; Weller, 1981). Ground nesters
and birds nesting on floating vegetation are the most
susceptible to damage, but even blackbirds and wrens with
nests attached to upright plants are subject to losses
in more severe storms (McNicholl, 1979). Species nesting
on floating mats or on the ground are likely to relay
quickly if losses occur early enough in the nesting season
(Bennett, 1938; Sowls, 1955; Weller, 1981). In at least
some species, nests tend to be more elaborate in marshes
than those of closely related species that usually nest
in more stable habitat (McNicholl, 1982) and more stable
structures, such as muskrat houses or old grebe nests
may be used if available (McNicholl, 1982). Nest building
behavior may also reappear later in the nesting cycle
than in related species, or even in the same species (Burger,
1980), thus allowing repairs to damaged nests.

Species nesting in unstable or unpredictable habitats
appear to have a lower degree of site tenacity than those
nesting in highly stable habitats (McNicholl, 1975; Buckley
and Buckley, 1980), and in colonial species, group adherence--a
social attachment to one's neighbors--may be enhanced,
allowing rapid pioneering of newly suitable habitat, including
habitat that was previously suitable but temporarily too
deep or too shallow. In extreme cases, a species may
nest in habitat that is generally unsuitable, such as
the dry land gull and grebe colonies cited in McNicholl
(1975), or in long-lived species may not breed at all
until conditions become suitable again (Hosford, 1965;
Evans, 1972).

Marsh nesting species also show a considerable degree of opportunism in their diet and in their foraging methods (McNicholl, 1980, 1984; Orians, 1980).

As in other types of habitat, succession by marsh plants will result in changes of the avifauna present (Weller and Spatcher, 1965; McCracken, 1981; Weller, 1981), and without periodic flooding marshes will become choked with vegetation and eventually become dry land sites (Sowls, 1955; Green et al., 1964; Weller, 1981). Thus, again, without some fluctuations in water levels, birds of marsh habitat will lose the chance to nest at all, unless, like the red-winged blackbird, they undergo a major change in nesting requirements.

In short, species of birds nesting in the habitats that are most susceptible to damage from water level changes are also the species best adapted to coping with such problems.

MOULTING

Most birds undergo a gradual moult after breeding, but waterfowl and a few other groups become flightless, and thus highly vulnerable to predation. During this short period of the life cycle, some waterfowl congregate on large lakes or bays where they are are presumably safer (Sowls, 1955; Bergman, 1973; Hochbaum, 1981). Such moulting areas are generally too large to be affected by most water fluctuations, and in any case, I am not aware of any such sites on the Great Lakes.

MIGRATION STOP-OVERS AND FEEDING SITES

Although nest-site fidelity has long been known to biologists (Buckley and Buckley, 1980), migration and wintering site fidelity are less well documented (reviewed by Smith and Houghton, 1984). Some known migration stop-over sites are extremely important to particular species. For example, Senner and Norton (1976) reported that virtually the entire North American breeding populations of western sandpipers (Calidris mauri) and dunlin (C. alpina) feed along the extensive intertidal flats of the Copper River delta in Alaska each May. Similar examples are known for waterfowl. Fluctuations in water levels beyond normal changes could profoundly affect the species using such sites.

At Long Point, Ontario, Bradstreet et al. (1977) found a significant decline from 1966 to 1971 in numbers of migrant shorebirds using beach pools as feeding sites but did not detect a similar decline in species that fed at the interface of the beach and Lake Erie. They attributed the decline to a rise in lake levels, a factor that adversely affected the beach pools but did not affect the beach-lake interface.

WATERBIRDS WINTERING ON THE GREAT LAKES

Concentrations of gulls and waterfowl along the Niagara River in early winter are well known (Goodwin, 1982), and diving ducks, especially oldsquaw (_Clangula hyemalis_), can be seen in Toronto Harbor all winter (pers. obs.). In fact, waterfowl wintering in the Toronto area are believed to be increasing (Goodwin _et al._, 1977). Little is known of wintering populations of waterbirds on the Great Lakes in general, but studies by Lambert (1981) in the winters of 1979-1980 and 1980-1981 indicated that large numbers of both gulls and waterfowl winter on the Great Lakes, especially in the vicinity of the Detroit River and on the lower lakes. Away from areas where these birds depend on thermal effluents, changes in water levels could profoundly alter the food supply of these birds and the amount of open water allowing them to gain access to the food.

SUMMARY

In this brief review, I have discussed some of the known effects of water level fluctuations on birds. As with most aspects of avian biology, the topic has been most studied during the nesting season, and relatively little is known about effects on migrants or wintering birds. Species nesting in highly stable habitats will generally be less affected by fluctuations in water level, but those likely to be affected have evolved a wide array of adaptations to frequent change. Migration and wintering site fidelity may be more important than previously realized, and the suitability of sites could be profoundly influenced by changes in water levels.

RECOMMENDED RESEARCH

In spite of the fact that most is known about nesting birds, there are still relatively few studies of bird communities in wetlands over long periods of time, and long-term effects of water fluctuations on birds therein can be predicted only at a generalized level until such studies are done. The sort of study conducted by Weller and Spatcher (1965) and by McCracken (1981) should be carried out at more sites over longer periods of time. Surveys of colonial waterbirds on the Great Lakes have been conducted annually by the Canadian Wildlife Service and several American agencies and provided that these are maintained, we have a much better basis for monitoring these species.

At the species level, the basic breeding biology of most species of marsh bird remains to be sorted out, the red-winged blackbird and some ducks being the only species for which many studies are available to date. Of the colonial waterbirds, the herring and ring-billed gulls are the best studied, and much of the research has been done on Great Lakes colonies by Blokpoel, the Ludwigs, Ryder, the Southerns, Weseloh and others. Common terns have also received considerable study on the Great Lakes by Blokpoel, the Ludwigs, and Morris, among others, but most of the other species have not been studied in this area in detail.

Should water level fluctuations be found to be affecting birds in the Great Lakes area adversely, experiments on remedial measures such as nest platforms (Sutcliffe, 1979) would be advised. The only such management experiments on the Great Lakes of which I am aware is of platforms for Forster's terns in the Green Bay marshes (J. Trick, pers. comm.). Water level manipulations could also be tried, but unless the importance of periodic flooding of at least some types of wetland is realized, such manipulations may be risky at best.

Finally, much more needs to be done in terms of basic survey work to determine locations and importance of concentrations of migrants and wintering birds in the Great Lakes area before any serious attempt can be made to determine effects of changes in water levels on them. With proposals to open the Great Lakes to year-round navigation and to divert Great Lakes water elsewhere, such studies may become urgent.

DISCUSSION

Donald L. Beaver
Department of Zoology and The Museum
Michigan State University
East Lansing, Michigan 48824

McNicholl has outlined and reviewed the main features of the effects of water fluctuations in the Great Lakes on avian populations using wetlands for feeding, nesting or other activities. The picture is one of dynamic change.

Habitats range from stable, very slow changing to ones with rapid and dramatic changes within even one season. The rates of change are affected by changes in water level, storms, the normal processes of erosion and vegetative growth and die-off. Avian populations using Great Lakes habitats do not show uniform response to these factors. Some species may be favored by one set of changes, such as flooding that kills vegetation which allows nesting, whereas others will be adversely affected. In sum, there are a diversity of possible responses.

As one of the goals of this organization is to understand the nature of these processes as they affect particular species groups, we must ask the question: Can we understand these complexities in relation to the long-term population dynamics of single species or groups of species using the Great Lakes wetlands? Furthermore, and perhaps even more importantly, can an understanding be gained that will allow predictions of the effects of water dynamics on avian populations?

McNicholl proposes that further understanding can be gained from more extensive research, especially broad surveys of single species, or groups of species utilizing Great Lakes wetlands. While initial surveys are necessary to document a species' current use of, and status in, a habitat, further surveys will only serve to document change or lack thereof in the populations under study. In other words, we are able to record changes or lack of them, but we often do not know why what was observed occurred nor what determined the magnitude of it. The value of long-term surveys, and perhaps other modes of study as well, will be seriously compromised in this research atmosphere unless an additional element is added. This element is an hypothesis, or model, of the proposed influence of water dynamics on avian populations as mediated through habitat use, direct influence, or both. The goal of such an hypothesis, or hypotheses, should be to predict short- and long-term trends in avian population dynamics based on our current understanding of the mechanisms at work. The literature is abundant with studies of species using wetlands in other regions for feeding, nesting and other activities as cited by McNicholl. Studies by Weller (1981) and Weller and Spatcher (1965) provide especially good starting points. There is already a base of information for Great Lakes species, as McNicholl has presented. Will at least some of this information be relevant to problems considered here? It seems likely, since there is little reason to expect the processes of flooding, erosion, vegetation growth, and so on to be different in their main effects in similar habitats with the same species elsewhere. However, even if little existing research is relevant to the problems in the Great Lakes, the process of building a research model will focus our thinking on the critical data needed, and the relationships that must be understood before predictions can be made with any level of confidence. These studies would receive top priority in future research efforts.

Some of the essential elements of a predictive model for the study of water dynamics and avian populations in wetlands are outlined by McNicholl and others already. We need to know the patterns of water level changes (both the amounts and the duration), rates of erosion of land forms holding habitat for wetland species, and the response of vegetation to these changes. Vegetation responses include rates of growth or die-off, alteration in pattern of dispersion and general vigor. Avian responses to these factors would include habitat selection for nesting, feeding and resting, mortality and natality responses such as ability to withstand flooding or drying, and ability to renest after destruction of a nest. Other important relationships would probably emerge with further study.

The benefits of this approach are that our needs for further research will have been focused. Future studies, such as the long-term surveys mentioned by McNicholl, can be compared against predictions. Understanding of the system will thereby be advanced in a more orderly and expeditious way. For many of the Great Lakes wetlands, our efforts are already too late. Time is short for what remains. Our studies should be as directed and timely as we can make them.

LITERATURE CITED

Bennett, L.J. 1938. The Blue-Winged Teal; Its Ecology and Management. Collegiate Press, Ames, xiv + 144 pp.

Bergman, R.D. 1973. Use of southern boreal lakes by post breeding canvasbacks and redheads. J. Wildl. Mange. 37:160-170.

Blokpoel, H. 1977. Gulls and terns nesting in northern Lake Ontario and the upper St. Lawrence River. Can. Wildl. Serv. Progress Notes No. 75. 12 pp.

Bradstreet, M.S.W., W.G. Page, and W.G. Johnston. 1977. Shorebirds at Long Point, Lake Erie, 1966-1971: Seasonal occurrence, habitat preference and variation in abundance. Can. Field-Nat. 91:225-236.

Buckley, F.G. and P.A. Buckley. 1980. Habitat selection and marine birds. In: J. Burger, B.L. Olla, and H.E. Winn (eds.), Behaviour of Marine Animals, Vol. 4, Marine Birds. Plenum Press, New York, pp. 69-112.

Burger, J. 1980. Nesting adaptation of herring gull (Larus argentatus) to salt marshes and storm tides. Biol. of Behav. 5:147-162.

Clapp, R.B., D. Morgan-Jacobs, and R.C. Banks. 1983. Marine birds of the southeastern United States and Gulf of Mexico, Part III. Charadriiformes. U.S. Dept. Interior, Washington, FWS/OBS-83/30, xvi + 854 pp.

Courtenay, P.A. and H. Blokpoel. 1983. Distribution and numbers of common terns on the lower Great Lakes during 1900-1980: A review. Colonial Waterbirds. 6:107-120.

Evans, R.M. 1972. Some effects of water level on reproductive success of the white pelican at East Shoal Lake, Manitoba. Can. Field-Nat. 86:151-153.

Goodwin, C.E. 1982. A Bird-Finding Guide to Ontario. Univ. Toronto Press, Toronto. 248 pp.

Goodwin, C.E., W. Freedman, and S.M. McKay. 1977. Population trends in waterfowl wintering in the Toronto region, 1929 to 1976. Ont. Field Biol. 31(2):1-28.

Green, W.E., L.G. MacNamara, and F.M. Uhler. 1964. Water off and on. In: J.P. Linduska (ed.), Waterfowl Tomorrow. U.S. Dept. Interior, Fish and Wildl. Serv., Washington. pp. 557-568.

Hochbaum, H.A. 1981. The Canvasback on a Prairie Marsh, Third Edition. Univ. Nebraska Press, Lincoln, xxii + 207 pp.

Hosford, H. 1965. Breeding success of the white pelican in two colonies in Manitoba in 1964. Blue Jay. 23:21-24.

Krannitz, P. 1984. A breeding colony of black skimmers, large-billed terns and yellow-billed terns in the Amazon: Selective pressures of rising water on nest-site choice and incubation time? Colonial Waterbird Group 8th Ann. Meeting Program and Abstracts: 9.

Lambert, A. 1981. Assessment of effects of winter navigation on bird populations on the Great Lakes. Long Point Bird Observatory, Port Rowan, Ont. for U.S. Fish and Wildl. Serv., 105 pp.

Lynch, J.J. 1964. Weather. In: J.P. Linduska (ed.), Waterfowl Tomorrow. U.S. Dept. Interior, Fish and Wildl. Ser., Washington, pp. 283-292.

McCracken, J.D. 1981. Avifaunal surveys in the cattail marshes at Long Point. Can. Wildl. Serv., Ontario region, MS rept., iv + 59 pp.

McCracken, J.D., M.S.W. Bradstreet, and G.L. Holroyd. 1981. Breeding birds of Long Point, Lake Erie. Can. Wildl. Serv. Rept. Ser., No. 44, 74 pp.

McNicholl, M.K. 1975. Larid site tenacity and group adherence in relation to habitat. Auk. 92:98-104.

McNicholl, M.K. 1979. Destruction to nesting birds on a marsh bay by a single storm. Prairie Nat. 11:60-62.

NcNicholl, M.K. 1980. Territories of Forster's terns. Proc. 1979 Colonial Waterbird Group Conf. 3:196-203.

McNicholl, M.K. 1981. Fly-catching by male red-winged blackbords. Blue Jay. 39:206-207.

McNicholl, M.K. 1982. Factors affecting reproductive success of Forster's terns at Delta Marsh, Manitoba. Colonial Waterbirds. 5:32-38.

McNicholl, M.K. 1984. Food and feeding behaviour of Forster's terns. Colonial Waterbird Group 8th Ann. Meeting Program and Abstracts: 14.

Miller, R.S. 1968. Conditions of competition between redwings and yellowheaded blackbirds. J. Animal Ecol. 37:43-61.

Orians, G.H. 1980. Some Adaptations of Marsh-Nesting Blackbirds. Princeton Univ. Press, New Jersey, xii + 295 pp.

Peck, G.K. 1976. Recent revisions to the list of Ontario's breeding birds. Ont. Field Biol. 30(2):9-16.

Peck, G.K. and R.D. James. 1983. Breeding Birds of Ontario. Nidiology and Distribution, Vol. 1. Nonpasserines. Royal Ont. Mus., Toronto, xii + 321 pp.

Scharf, W.C. and G.W. Shugart. 1984. Distribution and phenology of nesting Forster's terns in eastern Lake Huron and Lake St. Clair. Wilson Bull., 96:306-309.

Senner, S. and D.W. Norton. 1976. Shorebird migration and oil development on the Copper River delta area. Proc. 27th Alaska Sci. Conf., Vol. 1:179.

Smith, P.W. and N.T. Houghton. 1984. Fidelity of semipalmated plovers to a migration stopover area. J. Field Ornithol. 55:247-249.

Sowls, L.K. 1955. Prairie Ducks. Wildl. Manage. Inst., Washington, 193 pp.

Sutcliffe, S.A. 1979. Artificial common loon nesting site construction, placement and utilization in New Hampshire. In: S.A. Sutcliffe (ed.), The Common Loon, Proceedings of the Second North America Conference on Common Loon Research and Management, Natl. Audubon Soc., New York, pp. 147-152.

Vermeer, K. and L. Rankin. 1984. Population trends in nesting double-crested and pelagic cormorants in Canada. Murrelet. 65:1-9.

Weller, M.W. 1981. Freshwater Marshes; Ecology and Wildlife Management. Univ. Minnesota Press, Minneapolis, xv + 146 pp.

Weller, M.W. and C.S. Spatcher. 1965. Role of Habitat in the Distribution and Abundance of Marsh Birds. Iowa State Univ. of Science and Technology, Agric. and Home Economics Exper. Sta. Spec. Rept. No. 43, 31 pp.

Weseloh, D.V., S.M. Teeple, and M. Gilbertson. 1983. Double-crested cormorants of the Great Lakes: Egg-laying parameters, reproductive failure, and contaminant residues in eggs, Lake Huron 1972-1973. Can. J. Zool. 61:427-436.

CHAPTER 7

AVIAN COMMUNITIES IN CONTROLLED AND UNCONTROLLED GREAT LAKES WETLANDS

Harold H. Prince
Department of Fisheries and Wildlife
Michigan State University
East Lansing, Michigan 48824

ABSTRACT

Avian activities on four 47 ha to 200 ha wetland study areas were monitored over a four-year period. Two of the areas were diked so water levels could be controlled while the other two were subject to natural water level fluctuations. Nests of twenty species of birds were located in the study areas with eight species being well distributed. Red-winged blackbirds (Agelaius phoenicesus) and marsh wren (Cistothorus palustris) were the most common species. Both nest density and number of species increased as the percentage of open water decreased in the wetlands. Wetland study areas with poorly developed communities of submersed plants did not have as many species nesting and had more herons present in late summer compared to areas with well developed submersed plant communities. Rails responded to taped calls throughout the summer, and this technique may be useful for evaluating abundance of birds and productivity.

INTRODUCTION

Coastal wetlands in the Great Lakes region have declined both in quantity and quality. This prompts a need to understand and protect remaining areas before a permanent loss of a community or some of its members occurs. Our understanding of processes and species requirements in coastal wetlands of the Great Lakes is limited. Mudroch (1981) evaluated the nutrient dynamics of a Lake Erie wetland while Harris et al. (1981, 1983) has studied avian communities in wetlands of the Green Bay region. Other inferences about coastal wetlands must be made from studies of salt and freshwater wetlands.

Evolution in avian communities has been in response to vegetation and vegetative changes (Hilden, 1965). This becomes very apparent in unstable habitats such as wetlands where dramatic changes in structure often occur. A pattern of response of avian communities that relate to four identifiable hydrological stages in midwest glacial marshes has been proposed (Weller and Spatcher, 1965; Weller and Fredrickson, 1974; Weller, 1978). Changes in the avian community relative to wetland vegetative structure in the Great Lakes region has been described by Harris et al. (1977, 1981).

Coastal wetlands in the Great Lakes region provide a number of important functions to certain avian species during the ice-free period of the year. Habitat is provided for breeding, feeding, escape from predators, protection from environmental extremes, molting sites and staging areas. Each of these needs are satisfied by a particular configuration of vegetation and open water areas which is a function of the hydrological conditions. Although natural cycles once met all these requirements of the avian community throughout the year, changes in water quality and quantity as a result of human perturbation has changed the character of many areas. This confounds the understanding of avian response to wetland areas and forms the basis of this study which is to describe the utilization of Great Lakes coastal wetlands by birds during the breeding and postbreeding periods in wetlands with and without water control.

METHODS

Design and Study Areas

Four wetland areas were selected to represent different types of water control and vegetative configurations. One site was Pentwater Marsh in Oceana County, Michigan (Figure 1), and three areas were on the Shiawassee National Wildlife Refuge near Saginaw, Michigan (Figure 2). Pentwater Marsh and River Marsh were subject to natural water level fluctuations, and Pool 2 and Pool 4 were diked which allowed for controlled water levels except during periods of extreme flooding (Table 1). The vegetative community was established on all wetlands except Pool 4 which had been converted from agricultural use to wetland three years prior to this study.

Table 1. Status of water control and vegetation on wetland study sites.

Wetland	Water Control	Vegetation
Pentwater	Natural	Established
Pool 2	Diked	Established
Pool 4	Diked	Developing
River Marsh	Natural	Established

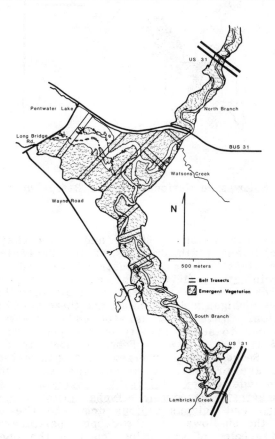

Figure 1. Pentwater Marsh in Oceana County, Michigan.

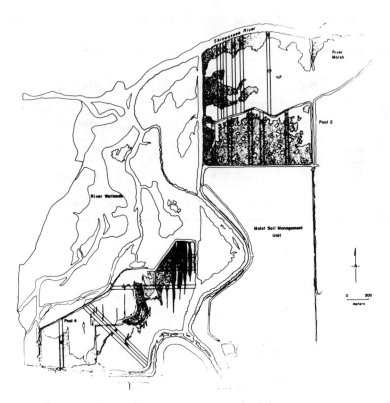

Figure 2. Shiawassee National Wildlife Refuge near Saginaw, Michigan.

       Pentwater Marsh is 86 ha and forms the junction of the North and South Branches of the Pentwater River. It can be classified as a palustrine wetlands according to Cowardin et al. (1979). The marsh is separated by a diked road with an unrestricted flow into Pentwater Lake. Lake Michigan is 4.5 km away, and there is no difference in elevation. Water flow is stopped or reversed every two hours due to seiche activity on the lake. Water level varied 0.5 m from a low in February, 1981, to a high in July, 1983. Sedge meadow, emergent, and open water habitat complexes are present. Dominant species in the sedge meadow include Carex stricta, C. rostrata, C. aquatilis, and Calamagrostis canadensis. Typha latifolia, Sparganium eurycarpum, and Scirpus validus dominate the emergent zone from the shallowest to the deepest part, respectively. Submersed species of vegetation found in the open water zone include, Potamogeton spp., Ceratophyllum spp., Utricularia spp., Myriophyllum spp., Cladophora spp. and Elodea spp. Lemna spp. and Najas spp. were also present in the open water zone.

The Shiawassee National Wildlife Refuge is 30 km from Saginaw Bay along the Saginaw River. Pool 2 and Pool 4 are diked palustrine wetlands 47 ha and 87 ha, respectively. The River Marsh contains 200 ha of undiked palustrine wetlands along the Shiawassee River. Elevation of the river above Saginaw Bay is 0.3 m. In addition to annual spring flooding, water levels in the River Marsh fluctuate up to 1 m due to wind and seiche action on Saginaw Bay. All emergent zones of vegetation on wetlands at Shiawassee national Wildlife Refuge were dominated by Typha spp. The River Marsh contained some Scirpus fluviatilis in scattered pockets. No submersed species of vegetation were found in the open water zone of the River Marsh because of the high turbidity of the water. Pool 2 had some Alisma spp. and Sagittaria spp. along the diked edges and a full complement of submersed species in the open water zone similar to those listed for Pentwater Marsh. Pool 4 was in the third year of establishment at the beginning of the study and contained a greater variety of species in the emergent zone than the other study sites. In addition to Typha spp., other emergent species of vegetation included Sagittaria spp., Alisma spp., Lythrum salicaria, Eleocharis spp., and Nymphaea spp. The open water zone did not contain many species of vegetation. Submersed species of vegetation were restricted in their distribution due to the high turbidity of the water and recent establishment of the wetland.

A series of 24 belt transects (30 m wide) totaling approximately 11,000 m were established on a random basis on the four wetland study sites. Each wetland area had an equal amount of sample area. The belt transect then formed the sample base for vegetative measures, nest, flush and call counts.

Vegetation

Estimates of above-ground primary productivity of dominant species emergent vegetation were made each month. Stem densities were determined from ten 1 m^2 plots systematically placed along the edge of each belt transect. Ten cattail stems, one from each plot, were then clipped along each transect, dried at 100° C for 24 hours and weighed. Standing crop was estimated by multiplying mean monthly stem density by stem weight.

Table 2.   Periods when nest searches were conducted on study wetlands from 1980 to 1983.

Wetland	Search Period	1980	1981	1982	1983
Pentwater	Early		x		x
	Mid-season		x	x	x
	Late	x		x	x
Pool 2	Early		x		x
	Mid-season		x	x	x
	Late	x	x	x	x
Pool 4	Early				x
	Mid-season		x	x	x
	Late	x		x	x
River Marsh	Early		x		x
	Mid-season		x	x	x
	Late	x	x	x	x

The amount of interspersion between emergent vegetation and open water greater than 1 m was estimated for each wetland area.   A line intercept that measured the length of Typha spp., mudflat, and open water areas was established along the edge of each belt transect.   The number of open water areas per 100 m of transect was calculated by counting the total number of openings along the transects and dividing the total by the number of 100 m segments.   Ratios of emergent vegetative cover to open water areas were calculated from the percentage of open water of the total distance covered by each habitat component.

Avian Surveys

Nest Counts.   Three nest searches were attempted on each wetland per year.   An early period ran from the last week in April until mid-May.   A mid-season search occurred during the last week of May and the first two weeks of June.   The late nest search was conducted during the last ten days of June and the first ten days of July.

Each 30 m belt transect was divided into 5 m search zones.   Each search zone was traversed by one person responsible for finding as many nests as possible at all strata. Groupings of two or three persons usually worked together on a nest search. Nests found were recorded by species, status of eggs or young, height, nesting substrate, and location along the transect within each 100 m segment. Although all belt transects were supposed to have been searched for the three periods during the four-year study, a variety of factors limited the search to 33 of the 48 possible search periods (Table 2). Only one search was possible during the initial year of the study while flooding and nest chronology affected the schedule in 1981 and 1982.

Call Counts. Members of the family Rallidae are secretive enough that in addition to the nest search, a call count technique was used to test for presence and possible reproductive activity. Taped calls of sora (Sora porzana), Virginia rail (Rallus limicola), American coot (Fulica americana), and common moorhen (Gallinula chloropus) were played at stations 100 m apart along the transects between sunrise and 1100 hrs at two-week intervals from May through September. Counts were not made when the wind speed exceeded 16 kph. The recording of each call was played with the speakers aimed in opposite directions at each station. The direction and distance of all responses were recorded for a period of one minute after the tape was stopped.

Flush Counts. Biweekly counts of large wading birds (Ardeidae) and waterfowl (Anatidae) were made along each belt transect. The counts were made in conjunction with the call counts. The locations of all flushed birds were recorded on field maps.

RESULTS

The emergent vegetation was dominated by Typha spp. in the wetlands at Shiawassee National Wildlife Refuge. Two additional species, Sparganium eurycarpum and Scripus validus, were also present at Pentwater Marsh. Standing crop dry weight estimates varied between month and year (Table 3). The estimates were the highest in August and ranged from a low of 421 g/m^2 in 1981 at Pentwater Marsh to 1723 g/m^2 in 1983 at Pool 4. Although estimates of standing crop usually increased as the season progressed, flood conditions at Shiawassee in 1981 decreased standing crop in Pool 2 and Pool 4 during July. Standing crop estimates were lowest for Pentwater Marsh and highest in the newly established wetland, Pool 4, and the River Marsh.

Structural descriptions of the wetlands were made each year on the basis of a line intercept along each

Table 3. Monthly estimates of cattail standing crop (g/m^2) at Pentwater and Shiawassee National Wildlife Refuge from 1981 to 1983.

Wetland	-----June-----			-----July-----			----August----		
	1981	1982	1983	1981	1982	1983	1981	1982	1983
Pentwater	221	280	288	349	530	--	421	610	627
Pool 2	320	340	532	340	730	973	770	1010	1248
Pool 4	650	560	695	560	910	1213	1140	1120	1723
River Marsh	930	430	458	430	1030	1057	1240	860	1217

belt transect. The number of openings greater than 1 m in length and percentage of open water per each 100 m segment were calculated for each year of the study (Table 4). Pool 2 had the greatest number of openings and a low percentage of open water per 100 m along the transects. Pentwater Marsh had the same percentage of openings but there was less interspersion. The River Marsh and Pool 4 were more open than Pool 2 and Pentwater Marsh. The degree and frequency of interspersion were similar for River Marsh and Pool 4 each year.

Table 4. Mean number ($\pm$ SD) of openings and percent open water per 100 m on four wetland study areas by year.

Wetland	----1981-----		----1982-----		----1983-----	
	Number Openings	% Open	Number Openings	% Open	Number Openings	% Open
Pentwater	0.6±0.1	29	1.0±0.4	32	1.4±0.7	27
Pool 2	2.4±0.8	30	6.8±1.8	42	6.2±3.0	24
Pool 4	1.9±1.4	64	1.5±0.4	64	2.5±1.8	71
River Marsh	0.5±0.1	61	1.2±0.5	77	0.8±0.4	74

Table 5.   Number of nests and searches ( ) on study wetlands by year.

Wetland	1980	1981	1982	1983
Pentwater	123 (1)	119 (2)	128 (2)	127 (2)
Pool 2	24 (1)	20 (3)[a]	15 (2)[a]	97 (3)
Pool 4	15 (1)	25 (1)	72 (2)	44 (2)
River Marsh	24 (1)	33 (3)	30 (3)	21 (2)
TOTAL	186	197	235	289

[a]A diversity of water treatments occurred which makes the total number of nests low compared with other wetlands.

Well-developed communities of submersed vegetation were present in the open water portions of Pentwater Marsh and Pool 2. The high turbidity of the water in Pool 4 and River Marsh prevented establishment of submersed vegetative species.

A total of 32 nest searches over a four-year period resulted in the discovery of 907 bird nests (Table 5). This amounted to an average location rate of 28 nests per search. The most nests per unit effort (46) were found in a series of late season searches conducted in

Table 6.   Number of bird species nesting on study wetlands by year.

Wetland	1980	1981	1982	1983	Total Number of Species
Pentwater	11	12	9	10	13
Pool 2	16	11	9	10	18
Pool 4	5	7	7	8	9
River Marsh	3	3	2	3	5

Table 7. Bird species that were found nesting within belt transects on the wetland study areas during 1980 to 1983.

Species	Pentwater				Pool 2				Pool 4				River Marsh			
	1980	1981	1982	1983	1980	1981	1982	1983	1980	1981	1982	1983	1980	1981	1982	1983
Pied-billed Grebe								x		x						
Least Bittern		x	x	x	x	x				x	x	x				
Canada Goose		x	x	x	x	x	x	x	x		x	x				
Green-winged Teal							x									
Black Duck					x											
Mallard	x	x	x	x	x	x	x	x				x				x
Redhead					x											
Virginia Rail		x			x	x	x	x								
Sora	x	x		x	x	x							x			
Common Moorhen	x	x	x	x	x	x	x	x	x	x	x	x				

Table 7. (Continued.)

Species	Pentwater				Pool 2				Pool 4				River Marsh			
	1980	1981	1982	1983	1980	1981	1982	1983	1980	1981	1982	1983	1980	1981	1982	1983
American Coot	x	x		x	x	x	x	x	x	x	x	x				
Wilson's Phalarope	x															
Black Tern	x	x	x	x		x	x	x								
Marsh Wren	x	x	x	x	x	x	x	x	x	x	x	x	x	x	x	x
Common Yellowthroat	x		x													
Song Sparrow	x	x	x			x										
Swamp Sparrow	x	x	x	x												
Red-winged Blackbird	x	x	x	x	x	x	x	x	x	x	x	x	x	x	x	x
Yellow-headed Blackbird	x				x			x		x	x	x				
Common Grackle						x										

1980. The number of nests found when multiple searches were conducted ranged from 22 per search in 1981 to 29 per search in 1983. The most nests were present on Pentwater Marsh and the least were found on the River Marsh. A diversity of water treatments occurred in Pool 2 in 1981 and 1982 which resulted in a low number of nests.

The greatest number of species nesting were found on Pool 2 in 1980, while the least number of species were present on the River Marsh in 1982 (Table 6). After 1980, similar numbers of nesting species were present on Pentwater Marsh and Pool 2. An intermediate number was present on Pool 4, and the lowest number occurred on the River Marsh. Nests of 20 species were located on all wetlands over the four-year period (Table 7). Marsh wren and red-winged blackbird nests were found on all study areas in all years. Three other species, Canada goose (Branta canadensis), common moorhen and American coot, were present

Table 8. Species that were found nesting along more than 20 percent of the 15 belt transect nest search zones for each wetland.

Wetland	Species	Number of Belt Transects
Pentwater	Mallard	3
	Black Tern	6
	Marsh Wren	12
	Red-winged Blackbird	15
	Swamp Sparrow[a]	8
Pool 2	Mallard	7
	Virginia Rail[a]	3
	Common Moorhen[a]	10
	American Coot	9
	Black Tern	6
	Marsh Wren	7
	Red-winged Blackbird	13
Pool 4	American Coot	7
	Marsh Wren	10
	Red-winged Blackbird	13
River Marsh	Marsh Wren	6
	Red-winged Blackbird	13

[a]Wide distribution is unique to wetland.

Table 9. The percentage of each transect open and the number of openings greater than 1 m on each transect in relation to the number of species and density of nests located by nest search date determined by multiple regressions.[a]

| | ---Early May--- | | --Early June--- | | ---Late June--- | |
| | Number | | Number | | Number | |
Variable	Species	Density	Species	Density	Species	Density
Percent Open						
P	0.224	0.069	0.002	0.001	0.020	0.001
R^2	0.025	0.055	0.151	0.168	0.089	0.161
Number Openings						
P	0.292	0.494	0.613	0.605	0.001	0.660
R^2	0.019	0.008	0.004	0.004	0.218	0.003

[a]Values in the table are the probability and respective R^2 for each variable.

on Pentwater Marsh, Pool 2 and Pool 4 in all years. Pentwater Marsh and Pool 2 also had nesting mallard (*Anas platyrhynchos*) and black tern (*Chlidonias niger*) in common. The remaining 13 species occurred sporadically over time or were unique to a specific wetland. Eight species of birds had nests widely distributed over one or more of the wetland areas (Table 8). Pool 4 and River Marsh contained the lowest number of species. Pool 2 contained seven species and two that were unique, Virginia rail and common moorhen. Swamp sparrows (*Melospiza georgiana*) nesting at Pentwater Marsh were unique to that wetland community.

The percent open and number of openings per 100 m of transect was compared to the number of species and density of nests on that particular segment for a three-year period (Table 9). There was a significant relationship on the basis of multiple regression between percent open and number of species nesting and nest density in early and late June. R^2 values ranged between 0.089 and 0.168. The number of openings per 100 m was significantly related to the number of species nesting for the late June period. The number of species found nesting and nest density was inversely related to the percent open per 100 m of transect

Table 10.   Mean number (± SD) of species and nest density in relation to percentage of open water.

% Open	—Number of Species—		—Nest Density—	
	Early June	Late June	Early June	Late June
0-9	--	--	--	--
10-19	3.1±1.7	2.8±1.1	9.6±8.4	11.5±10.4
20-29	3.0±1.3	3.3±1.8	6.3±1.2	4.3±2.4
30-39	2.6±1.1	2.3±1.8	11.1±13.6	4.3±4.0
40-49	--	--	--	--
50-59	1.4±1.1	2.2±1.6	1.6±1.5	2.0±2.2
60-69	3.0±3.4	2.7±2.8	4.0±2.6	2.7±2.1
70-79	1.7±1.2	2.0±1.6	3.0±3.2	2.0±1.6
80-89	1.7±1.4	1.3±0.5	1.7±0.8	1.4±0.5
90-100	--	--	--	--

Table 11.   Numbers of Rallidae by species responding to tape recorded calls in Pool 2 by month during 1981, 1982 and 1983.

Month	Year	Virginia Rail	Sora	Common Moorhen	American Coot
May	1981	12	2	3	0
	1982	5	0	0	0
	1983	--	--	--	--
June	1981	6	0	0	0
	1982	13	1	8	6
	1983	31	0	2	55
July	1981	32	15	1	3
	1982	45	16	10	5
	1983	78	4	6	46
August	1981	47	23	0	7
	1982	42	8	9	4
	1983	53	26	3	7
September	1981	11	54	0	6
	1982	19	30	0	0
	1983	--	--	--	--

(Table 10). An average of three species had between four
to twelve nests per ha at 10 to 39 percent open. This
dropped to two species with one to four nests per ha at
50 to 89 percent open.

Four species of rails responded to the taped calls
played at stations 100 m apart along each belt transect.
There were responses by all species, except American coot
in May, from May through September each year of the study
on Pool 2 (Table 11). Response of Virginia rail began
in May, increased in June and peaked in July and August.
Sora responses were low in May and June and increased
to a peak in September. Constant number of common moorhen
and American coot responded from June through August.

All four species of Rallidae were heard in all wetlands
on all years except for common moorhen and American coot
in the River Marsh (Table 12). The greatest number of
Virginia rail and sora responses were in Pool 2. American
coot, common moorhen, and Virginia rail usually responded
the most on Pentwater and Pool 4. Although all rail responses
were low in the River Marsh, Soras were the most common.

Table 12. Mean numbers of responses by Rallidae to tape
recorded calls on wetland study areas in 1981, 1982 and
1983.

Wetland	Year	Virginia Rail	Sora	Common Moorhen	American Coot
Pentwater	1981	4.3	1.0	2.8	5.3
	1982	1.5	1.0	3.0	0
	1983	2.6	0.6	1.8	3.0
Pool 2	1981	21.6	18.8	0.8	3.2
	1982	24.8	11.0	5.4	3.0
	1983	32.4	6.0	2.2	21.6
Pool 4	1981	1.7	0.3	0.3	13.3
	1982	2.7	6.3	2.7	4.7
	1983	3.8	1.4	0.6	6.0
River Marsh	1981	2.8	6.5	0	0
	1982	1.3	2.3	0	0
	1983	0	1.3	0	0

Table 13. Numbers of Ardeida observed on strip census routes in wetland study areas in 1981, 1982 and 1983.[a]

Species	Year	Pool 2			Pool 4			River Marsh		
		15 July-1 Aug.	-15 Aug.	-30 Aug.	15 July-1 Aug.	-15 Aug.	-30 Aug.	15 July-15 Aug.	-15 Aug.	-30 Aug.
Great-Blue Heron	1981	7	2	4	7	10	6	3	3	9
	1982	4	4	0	7	2	1	5	4	18
	1983	2	3	6	12	5	14	1	7	2
Black-Crowned Night Heron	1981	9	—	—	—	—	9	—	1	12
	1982	1	5	—	11	18	1	—	4	2
	1983	—	2	4	7	15	19	—	1	11
Great Egret	1981	—	—	—	13	22	—	—	—	—
	1982	—	3	—	17	14	32	—	2	21
	1983	—	2	—	8	32	11	—	1	12
Least Bittern	1981	—	—	—	—	—	1	—	—	—
	1982	—	—	—	1	—	—	—	—	—
	1983	—	—	—	1	—	—	—	—	—
Green-Backed Heron	1981	1	—	—	—	—	3	2	—	—
	1982	—	—	1	—	—	3	—	—	—
	1983	—	2	1	—	3	—	—	—	—

[a]Herons were not observed on transects at Pentwater. Up to three great-blue herons and five least bitterns were observed there during regular censuses.

Great-blue heron (Ardea herodias) was the most common member of the Family Ardeidae observed on the strip census made along each belt transect (Table 13). Black-crowned night heron (Nycticorax nycticorax) were commonly observed on the Shiawassee wetland but not at Pentwater. Great egrets (Casmerodius albus) were present on Pool 4 and River Marsh. Low numbers of least bittern (Ixobrychus exilis) and green-backed heron (Butorides striatus) were observed on the wetlands. Pool 4 and River Marsh had the greatest variety and number of herons in late summer of each year. Herons were rarely observed at Pentwater Marsh with the exception of least bitterns.

DISCUSSION

All wetland study areas were classified as Palustrine Systems according to the wetland classification of Cowardin et al. (1979) or as semipermanent according to Stewart and Kantrud (1971). All of the areas contained persistent emergent plants in relatively calm water. Pentwater Marsh and Pool 2 had well-developed submersed plant communities while Pool 4 and River Marsh did not. Pentwater Marsh was also influenced by a sedge meadow complex and a shrub-carr zone and was very similar to Green Bay wetlands described by Harris et al. (1977). The distribution and amounts of these vegetational assemblages directly influenced the use of each area by birds.

The major activities of birds using wetland areas were for feeding, nesting, resting, and escape. The study areas were selected to provide comparisons of avian responses throughout the months of May through September. Birds were present on all areas throughout the entire period of observation.

The placement of nests in a marsh occurred in one of three strata that have been described by Weller and Fredrickson (1974). The strata include nests located on the water level on floating vegetation or in emergents, above the water level in short and weak vegetation like sedges, or above the water level in tall cattail and similar robust emergents. Twenty species of birds used the wetland study areas for nesting. Eight species were well distributed (> 20% of the transects) in at least one study area, and marsh wrens and red-winged blackbirds were the only species common to all areas. Marsh wrens and red-winged blackbirds nest in the strata where short and weak vegetation occur (Weller and Fredrickson, 1974), and this is the only component in all wetlands that was abundant enough to be commonly used. Six common species used the nest strata at the water level where nests are built in emergent vegetation, and black tern used the strata at water level where nests are placed on floating debris. The tall cattail and robust emergent strata was not widely distributed enough to have any common nesting species associated with it.

Horizontal zones of vegetation have also been related to zonation of nesting birds in wetlands (Weller and Fredrickson, 1974). The peripheral trees and shrubs and peripheral marsh emergents were used by red-winged blackbirds in all study sites. Marsh wren were distributed in the peripheral marsh emergent zone in all study sites. The species associated with the central robust emergent zone and open water zone were only found in Pool 2 and Pentwater Marsh.

The number of species nesting and nest density were inversely related to percent of open water along each transect. This seems to be reverse of the trend found by Weller and Fredrickson (1974). The two wetland study areas that were most open, Pool 4 and River Marsh, did not have a well-developed submersed plant community in the open water zone because of high turbidity in the water. This eliminated the species of birds that nest at the water level on debris and a potential food source for bird species that nest in other strata. Pool 2 and the River Marsh were located adjacent to each other. The water was clear in Pool 2 as a result of being impounded, and there was a well-developed submersed plant community in the open water zone. The highly turbid river water prevented the growth of submersed species of aquatic plants which resulted in the least diverse community of nesting birds occurring next to the most diverse community. Muskrat (Ondatra zibethicus) activity was high in Pool 2 and very limited in the River Marsh. Although many studies have identified vegetative structure as an important variable influencing utilization of wetlands by nesting birds, most investigations did not make a comparison of areas with and without submersed vegetation. The value of the submersed community as a cue seems to have been underrated.

The response of rails to taped calls appeared to give a good indication of relative abundance. The high response of Virginia rails in Pool 2 and American coot in Pool 2 and Pool 4 were consistent with the wide distribution (> 20%) of nests in the areas. The large number of American coots on Pool 4 is consistent with Weller and Fredrickson's (1974) observations that this species pioneers readily to new areas and they are often followed or replaced by common moorhens over time. Kantrud (1985) reports that American coots use the semipermanent wetlands with well-developed deep marsh zones in North Dakota.

The increase in response of sora and Virginia rails in August and September were different than American coot and common moorhen. The increase in calling frequency could have been indicative of the yearly production, with the young birds beginning to respond, or a premigratory buildup on Pool 2 or both. Some of the responses in August and September definitely sounded like birds that were beginning to call for the first time.

Although the River Marsh and Pool 4 were least used as nest areas among the four wetland study areas, these areas were used most frequently by the herons during the mid to late summer. Great-blue herons and great egrets were most often observed. Both wetland areas had poorly developed vegetative communities. About two-thirds of these areas were open water, and the submersed plant species were not present because of the high turbidity of the water.

Many evaluation systems of wetland quality place a high priority on the value of an area for nesting species. Although this is important, this study points out that other environmental conditions will prompt considerable use by different species. It is very important that data be collected on avian activities throughout the ice-free period in order to evaluate the importance of wetland habitats.

The dynamic nature of wetland habitats has been well established by most investigators. This is in response to the variety of hydrological regimes that can and do occur. The high mobility of the avian community make them well adapted to take advantage of the habitats. Different wetland environments in this study prompted a variety of avian responses. It appeared that species were just waiting for the proper environments in order to respond. Where is the reservoir of individuals for such a response? This was not a problem for most species when wetland habitats were well distributed. Now that wetland habitats have decreased in size and diversity, knowledge about these factors becomes important for our understanding of community interactions.

ACKNOWLEDGMENTS

I thank Douglas Reeves who supervised this study and participated in the field work during the first three years and Rick Rusz who directed the field research during the fourth year. Thanks are due to the many field assistants for their help. The staff at the Shiawassee National Wildlife Refuge were very helpful. James Kelley and Thomas

Burton kindly provided hydrological and vegetative data for Pentwater Marsh. Thanks are due to David Gordon for assistance with analysis of data. I thank the Michigan Sea Grant College Program and Michigan Agricultural Experiment Station for financial support. This is Michigan Agricultural Experiment Station Journal 11698.

LITERATURE CITED

Cowardin, L.M., V. Carter, F.C. Golet, and E.T. LaRoe. 1979. Classification of wetlands and deepwater habitats of the United States. U.S. Fish and Wildl. Serv. Biol. Sur. Prog. FWS/OBS-79/31., 103 pp.

Harris, H.J., T.R. Bosley, and F.D. Roznik. 1977. Green Bay's coastal wetlands--a picture of dynamic change. pp. 337-357, in: Proceedings of Waubesa Conference on Wetlands, Madison, Wisconsin.

Harris, H.J., G. Fewless, M. Milligan, and W. Johnson. 1981. Recovery processes and habitat quality in a freshwater coastal marsh following a natural disturbance. Proc. Midwest Conf. on Wetland Values and Management, St. Paul, Minnesota.

Harris, H.J., M.S. Milligan, and G.A. Fewless. 1983. Diversity: Quantification and ecological evaluation in freshwater marshes. Biol. Cons. 27:99-110.

Hilden, O. 1965. Habitat selection in birds. Ann. Zool. Fenn. 2:53-75.

Kantrud, H.A. 1985. American coot habitat in North Dakota. Prairie Nat. 17:23-32.

Mudroch, A. 1981. A study of selected Great Lakes coastal marshes. National Water Research Institute, Scientific Series No. 122, 44 pp.

Stewart, R.E. and H.A. Kantrud. 1971. Classification of natural ponds and lakes in the glaciated prairie region. U.S. Fish and Wildl. Serv. Resour. Publ. 92.

Weller, M.W. 1978. Management of freshwater marshes for wildlife. Pp. 267-284, in: R.E. Good, D.F. Whigham, and R.L. Simpson (eds.), Freshwater Wetlands. Academic Press, NY.

Weller, M.W. and C.E. Spatcher. 1965. Role of habitat in the distribution and abundance of marsh birds. Iowa Agriculture and Home Economics Experiment Station Special Report No. 43.

Weller, M.W. and L.H. Fredrickson. 1974. Avian ecology
of a managed glacial marsh. Living Bird 12:269-291.

CHAPTER 8

RELATIONSHIPS OF WATER LEVEL FLUCTUATIONS AND FISH

Charles R. Liston and Saralee Chubb
Department of Fisheries and Wildlife
Michigan State University
East Lansing, Michigan 48824-1222

Alteration of water levels to attempt to manipulate fish populations and communities has been used extensively by fish management agencies on many of the nation's hydroelectric reservoirs, flood control reservoirs, recreational fishing lakes, farm ponds, and irrigation reservoirs. A general working plan for southern reservoirs could be as follows: a full lake is maintained during the spring spawning season and during the summer growing season; in early fall, after production of fish food is nearly complete (invertebrates and small forage fishes), water levels are lowered ("the fall drawdown") which will concentrate organisms; predator species such as bass, crappie, and catfishes then can consume their "easy prey," thus growing faster and larger, and desirable thinning of populations of small species such as bluegill can occur; the reservoir is refilled with spring rains and a successful spawn of desirable species may occur, partially because the small fishes that prey on eggs and young were much reduced in numbers over fall and winter; the frequency of the drawdown (annual, semi-annual, etc.) will depend upon the type of reservoir and management goals (Hulsey, 1958).

The above water level management scheme reflects somewhat the natural changes of hydrologic conditions in most of our lakes and streams, i.e., increased spring rainfall, followed by late summer and fall decreased rainfall which creates seasonal increasing and decreasing water levels. Abnormally high water levels during spring may have significant effects. Keith (1975) outlined some of these as follows:

(1)   shoreline terrestrial vegetation is flooded which initiates dying and decomposition and subsequent release of nutrients, thus increasing the water productivity;

(2)   fish food organisms such as insects and earthworms are quickly added to the water;

(3)   new cover and habitat for shoreline fish species is added; and

(4)   an area of water is created that is sparsely populated with fish, which should stimulate reproduction and growth as fish attempt to fill the "void."

Certain species of fish, especially largemouth bass, do best when water level increases occur immediately before, during, and for a short time following the spawning and nursery period. Though long-term data on standing stocks of fish in relation to changing water levels are rare, especially in the Great Lakes area, some data from reservoirs appear to show direct benefits of high water levels regarding production of young-of-the-year (YOY) fish. Data from Keith (1975) on Bull Shoals Reservoir, Arkansas, are an example. Long-term sampling in coves with rotenone (1954 through 1974) showed that during years when water level elevations were abnormally high during spawning and nursery periods, abundances of YOY bass were also high compared to years of lower water levels (Figures 1 and 2). Survivorship and growth of the large crops of YOY bass are apparently directly benefited by high water levels also, because high water levels stimulate spawning of many small forage species including minnows, sunfishes, and gizzard shad (Keith, 1975). In a more recent study on largemouth Bass, Miranda et al. (1984) observed that by raising the water level above normal summer pond in an Alabama-Georgia reservoir early survival of YOY was enhanced, and that survival and abundance of YOY were both positively related to water level. Other important species exhibit a similar dependency on water levels. Studies of both reservoirs and inland marshes have documented a marked reduction of northern pike year-class strength with reduced water levels (Hassler, 1970; Carbine, 1943). Chevalier (1977) reported that lake level at spawning time was important for walleye spawning success.

Beam (1983) in a five-year study of water level management and white crappie population trends in a Kansas reservoir, concluded that increased water levels during spawning periods were significantly related to increases in year class strengths of crappie. However, fluctuations in water levels during spawning seasons had a negative relationship to crappie year class strengths. Thus, high water levels tend to enhance reproductive success only if they are stabilized during the critical spawning period. Beam's study was done to examine the results of implementing a water level management plan, indicated in Figure 3. Water levels were to be increased 2 m during spring, stimulating natural reproduction and flooding vegetation. This increase was to be maintained until mid-July, when flooded vegetation would protect some species of larval fish. The mid-summer drawdown to elevation 241 (Figure 3) would dewater 20 percent of the basin, allowing the fluctuation zone to revegetate, and increased compaction of silt deposits

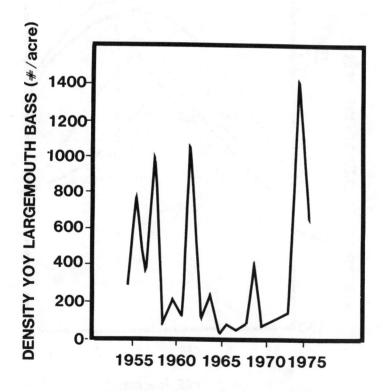

Figure 1. Number of young-of-the-year bass collected per acre with rotenone in coves on Bull Shoals Reservoir, Arkansas, from 1954 through 1974 (after Keith, 1975).

would occur. Grasses and millet could be seeded on exposed shores at this time. The fall water level rise of 1 m would flood vegetation for waterfowl attraction, and the fall drawdown would protect shorelines from ice damage.

Low water levels at improper periods may have detrimental effects on fish populations. Lantz et al. (1964) observed that the failure of water levels in a Louisiana lake to return to normal after a drawdown apparently caused the predatory game fish to fail to spawn. Detrimental effects from lowered water levels have been put to good use in managing rough fish populations such as carp. Tarzwell (1941) first pointed out the possibility of controlling

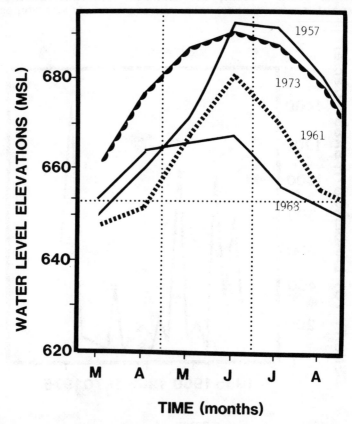

Figure 2. Water level elevations in Bull Shoals Reservoir, during four years when water levels were significantly above the normal power pool elevation during the fish spawning and nursery periods (after Keith, 1975).

carp in large reservoirs by water level manipulation. Shields (1958) found that planned water level drawdowns of 0.46 to 0.61 m in Fort Randall Reservoir, South Dakota, immediately following periods of major carp spawning over three years were primarily responsible for poor reproductions. Furthermore, winter drawdowns may control adult carp populations by restricting access to forage areas and/or by stranding carp in shallow areas where they may suffer increased mortality beneath the ice (McCrimmon, 1968). Thus, negative effects of dewatering areas have been turned into beneficial uses in some cases. Other potential beneficial effects of lowered water levels include control of obnoxious vegetation (Mathis, 1966), "conditioning" of organic materials allowing for nutrient release after reflooding, and reduction of stunted sunfishes and increase in size of game fishes (Lantz et al., 1964).

We have dealt thus far with data from outside the Great Lakes regarding water level fluctuations and fish community relationships. Very little is known of these relationships in Great Lakes coastal areas. It is generally not feasible to regulate water levels in our coastal marshes, the main exceptions being the controlled marshes along Western Lake Erie and the Lake St. Clair area. Lake Superior levels, though controlled since 1921 (International Great Lakes Levels Board, 1973), are managed primarily for power production and commerical shipping. Lake Ontario levels have been controlled since the late 1050s for power production

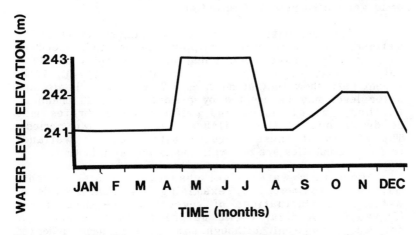

Figure 3. Monthly water levels established for a fisheries-oriented plan for annual water level management at Elk City Reservoir, Kansas (after Beam, 1983).

and shipping interests (U.S. Army Corps of Engineers, Buffalo District, personal communication). Great Lakes water levels tend to change gradually, being influenced by short-term and long-term climatic conditions, and exert their influence on levels of most connected wetlands generally independent of human intervention. Occasional steep barometric gradients may, however, influence brief changes (seiches) in water level (Hough, 1958), and winds may cause short-term significant changes in water level. For example, west winds on Lake Erie of 113-129 km/hr caused lake levels to differ by about 2.75 m at the two ends of the Lake on 30 April, 1984. Water levels at Toledo, Ohio, had dropped about 1.37 m below levels of the previous day, while near Buffalo, New York, levels were raised by a similar amount (Great Lakes Commission, 1984). Ice jams during early spring may occasionally create brief changes in water levels also, especially in connecting channels. For example, because of ice jams above Lake St. Clair in spring, 1984, water levels dropped more than 0.76 m in Lake St. Clair during 21 March to 19 April (Great Lakes Commission, 1984). Because of the extensive marshy areas along Lake St. Clair and because water levels fell at a critical time for northern pike and walleye spawning and nursery periods, this event may have detrimental effects on the 1984 year class strength of these species and perhaps others. However, although short-term events such as these may be quite noteworthy, their long-term impacts on fish communities in wetlands are probably much less than seasonal or longer period natural fluctuations.

Brief, repetitive water level changes in shoreline wetlands near commercial shipping lanes, influenced by passing ships, have been going on for decades but have only recently begun to be studied (McNabb et al., 1984). Recent data show that as much as a 70 cm change in wetland water level may be created by passing vessels in channels. Further, larval fishes and drifting invertebrates may be drawn out of the wetlands during drawdown periods. The effects of these frequent alterations of wetlands on fish communities are not well understood at this time.

Long-term water level changes in coastal wetlands may be approximated by examining historical records of water level fluctuations of several of the Great Lakes (Figure 4). A clear cyclic nature to these changes is not readily apparent, although some authors have indicated that in general a high water period and low water period may occur every seven to ten years (Jaworski et al., 1979). Maximum and minimum monthly lake levels since 1900 were as follows:

	High	Low	Range
Lake Michigan/			
Huron	581.1 (Jul '74)	575.4 (Mar '64)	5.7
Lake St. Clair	576.2 (Jun '73)	569.9 (Jan '36)	6.3
Lake Erie	573.5 (Jun '73)	567.5 (Dec '34)	6.0

Thus, water levels in coastal wetlands on Lakes Michigan/Huron, St. Clair, and Erie may be expected to fluctuate some 1.83 m over long periods of time. Smaller, seasonal variations occur annually. In general, water levels tend to be lowest during January/February, begin increasing during early spring to reach highs in July/August, then decline through late summer and fall (GLBC, 1975). The maximum (1860 - 1970) winter low to summer high change in water levels (m) has been as follows: Lake Superior, 0.58; Lake Michigan/Huron, 0.67; Lake St. Clair, 1.01; Lake Erie, 0.82; Lake Ontario, 1.07. The minimum (1860 - 1970) winter low to summer high change in water levels (m) has been: Lake Superior, 0.12; Lake Michigan/Huron, 0.03; Lake St. Clair, 0.27; Lake Erie, 0.15; Lake Ontario, 0.21 (GLBC, 1975).

The direct effects these natural water level changes will have on Great Lakes coastal wetlands will vary greatly

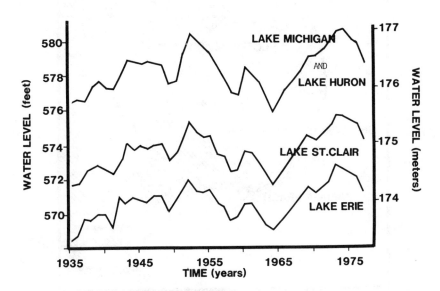

Figure 4. Lake level elevations of Lakes Michigan, Huron, St. Clair and Erie from 1935 through 1978 (Source: U.S. Department of Commerce, 1976).

with wetland type. Each wetland may be a unique environment, though the International Lake Erie Regulation Study Board (ILERSB, 1981) attempted to portray Great Lakes coastal wetlands as follows: unrestricted bay, restricted riverine, open shoreline, connected inland lake, shallow sloping beach, barrier beach, river delta, and man-made. These wetland types are graphically shown in Figure 5, as modified from Busch and Lewis, 1982. Primary effects of changing water levels such as changes in flow regimes, velocity, flushing rates, volume, current patterns, sedimentation,

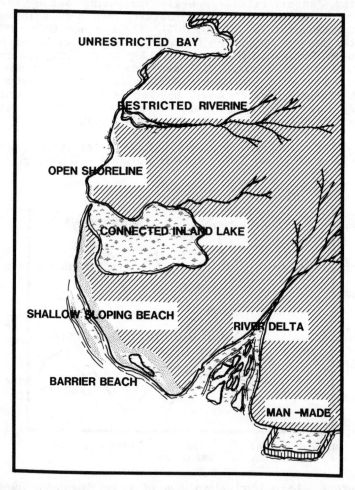

Figure 5. Pictoral representation of seven wetland types (modified from ILERSB, 1981).

temperature, dissolved oxygen, and concentration of dissolved and suspended solids would differ among these wetland types, creating different responses in vegetational patterns. We know of no long-term study on fish ecology and population dynamics within any of these wetland types in the Great Lakes, especially as related to changing water levels. Information on the Great Lakes littoral enviornment, though increasing in the past decade, is quite limited reflecting the traditional research emphasis on the deep water environment.

It is at best difficult to relate fish communities to changing water levels when the primary fisheries information is lacking. Busch and Lewis (1982) indicated that basic questions such as what fish use the Great Lakes wetlands, when they use them, and for what purposes still need to be answered. A recent survey by the U.S. Fish and Wildlife Service on fish resources in coastal Great Lakes marshes frequently refers to the lack of fish data (Herdendorf et al., 1981). Recent work on littoral zones of Lake Michigan (Brazo and Liston, 1979; Brazo, 1984; Liston and Chubb, 1983; Chubb, 1984; Jude et al., 1975; Jude et al., 1980; Liston et al., 1981) has helped develop quantitative data sets on larval, juvenile and adult fish populations and communities that could be used as baselines for future comparisons of littoral zone fish communities under different water levels regimes. Extensive data are now available on fish communities associated with wetlands of the St. Marys River, under water levels existing during 1979-1984 (Liston et al., 1983; 1984). Likewise, extensive data on fish communities in littoral zones of the St. Clair River-Lake St. Clair area are now being gathered by the Great Lakes Fisheries Laboratory, USFWS. On-going work by Johnson and Barnes (1983) on fish communities in controlled Lake Erie marshes will help future understanding of these habitats, and Whillans (1979) has examined changes in fish communities of bays in Lake Ontario. The USFWS (1982) identified 20 species of fish using a marsh on Lake Ontario. Work by Ringler et al. (1984) is leading to a better understanding of littoral zone fishes in the St. Lawrence River. Many of these data sets, when quantitative, can be extremely useful in determining effects of future changes of water levels in addition to changes in other lake characteristics in the years ahead. However, no significant experiments or long-term data collecting coupled with water level changes, either natural or manipulated, have been performed in the Great Lakes.

Species of fish within a wetland community must, of course, be known before any further steps can be taken towards understanding how wetland changes may affect the community. The species and relative abundances no doubt differ among wetland types (Figure 5) and according to the Great Lake with which they are associated. Our studies on Pentwater Marsh on Lake Michigan, a "restricted riverine" type, show a high diversity of 44 fish species, though the bulk of the juvenile and adult fish community is comprised of eight species including white sucker, pumpkinseed sunfish, largemouth bass, yellow perch, northern pike, black crappie, brown bullhead, and golden shiner (Table 1). With the exception of the white sucker, these major species spawn in or near emergent or submerged vegetation in shallow portions of the marsh, thus high water levels during spring/summer are needed to maintain spawning and nursery sites. Interestingly, of all major species the white sucker appears to be the only important Lake Michigan fish to also heavily use

Table 1. Species and numbers of juvenile and adult fish collected in Pentwater Marsh on Lake Michigan during 3/15/83 to 10/25/83. Gears included large mesh trap nets (52), small mesh trap nets (28), electroshocking (56) and experimental gill nets (12).

Species	Total No.	% TN	Species	Total No.	% TN
White sucker	2,058	24.9	Johnny darter	12	0.2
Pumpkinseed	929	11.2	Alewife	8	0.1
Largemouth bass	836	10.1	Brown trout	7	0.1
Yellow perch	802	9.7	Trout-perch	6	0.1
Northern pike	794	9.6	Rosyface shiner	6	0.1
Black crappie	709	8.6	Brook silverside	6	0.1
Brown bullhead	570	6.9	Chinook salmon	5	0.1
Golden shiner	>357	4.3	Mottled sculpin	5	0.1
Bowfin	247	3.0	Quillback carpsucker	4	0.1
Central mudminnow	157	1.9	Channel catfish	4	0.1
Bluntnose minnow	>133	1.6	Yellow bullhead	4	0.1
Silver redhorse	89	1.1	Lake trout	3	<0.1
Shorthead redhorse	84	1.0	Longnose gar	3	<0.1
Rock bass	84	1.0	Longnose sucker	2	<0.1
Rainbow trout	62	0.8	Walleye	2	<0.1
Carp	60	0.7	Hornyhead chub	2	<0.1
Coho salmon	47	0.6	Blackside darter	2	<0.1
Bluegill	45	0.5	Lake sturgeon	1	<0.1
Smallmouth bass	44	0.5	Freshwater drum	1	<0.1
Spottail shiner	32	0.4	Common shiner	1	<0.1
Golden redhorse	26	0.3	Mimic shiner	1	<0.1
Gizzard shad	14	0.2	Rainbow smelt	1	<0.1

the Pentwater Marsh throughout the year. The invertebrate food sources of this marsh are apparently freely available to this species, as well as food sources out in Lake Michigan.

It is hypothesized that not only high but stable spring/early summer water levels are important to the Pentwater fish community, as studies from reservoirs have indicated that production of YOY sunfish species is negatively affected when water levels fluctuate during the spawning/nursery periods. This should also be true for nothern pike, a species spawning in the shallowest, most vegetated portion of the marsh. Nelson (1978) noted that the decline of northern pike and yellow perch in Lake Oahe, a Missouri River reservoir, could be related to reduction of vegetational spawning habitat caused by fluctuating water levels. However, post-larval northern pike may require a current flow to induce the needed exodus from the upper marsh (Forney, 1968). Thus, although rapid fluctuation in water level may be detrimental, a gradual decrease may actually improve the survival of some post-larval fishes such as northern pike.

Unstable, fluctuating water levels may also alter the composition of benthic macroinvertebrates in littoral zones, favoring oligochaetes and chironomids over important prey groups as gastropods, amphipods, stoneflies, mayflies and caddisflies. Such changes may account for some of the lower productivities of sunfishes observed by other authors, and may be attributable to changes in substrate, specifically to the accumulation of silt and loss of vascular macrophytes (Hildebrand et al., 1980).

We suggest that maintenance of relatively high, stable water levels during post-spawning periods is important for production of food for larval fishes, especially when they begin feeding exogenously after yolk sac reabsorption. Growth of attached algae, fungi and associated animals on stems and leaves of plants may amount to 100-500 g dry wt/m^2 during some periods of the year (Westlake, 1966). Production of periphyton on stems and surfaces of littoral aquatic vegetation may supply significant food for micro-grazing invertebrates which in turn may be eaten by larval fish. In fact, some investigations suggest that the "critical period" regarding survival of year classes may be in the larval stage when exogenous feeding begins (Braun, 1967; Cushing, 1974). Relatively small changes in wetland water levels during spring and summer may result in large changes in total surface area of plants available for production of periphyton. Recent insight into this phenomena has been gained through studies of periphyton productivity in wetlands of the St. Marys River (McNabb, 1984). McNabb

studied the net productivity of periphyton on live and
dead stems of hard-stem bulrush (Scirpus acutus) and bur
reed (Sparganium eurycarpum), and extrapolated the results
over a square meter of wetland under various water depths,
giving estimates of the areal periphyton productivity
(mg C/m^2 wetland/day). The effects of water depths on
periphyton productivity growing on high density stands
of hard-stem bulrush may be seen in Figure 6. This influence
is exerted through its effect on leaf and stem area per
unit area of wetland available for periphyton development.
McNabb concluded that either a change in water depth or
emergent stem density may significantly change periphyton
production. A potential loss of periphyton production
with lowered water levels should be considered in future
decisions regarding water diversion from the Great Lakes
(Manny, 1984).

Water level fluctuations may also alter temperature
regimes in littoral zones, thus influencing fish spawning
periods and rates of food production. In Kuybishev Reservoir,
Russia, inundation of shallow areas during spring floods
resulted in rapid warming, stimulating the spawning of
fishes that spawn over vegetation and spawning migration.
Disruption of water warming patterns and absence of proper

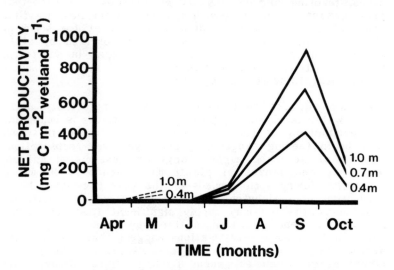

Figure 6. Areal net productivity of periphyton growing
on a high density stand of Scirpus acutus at three water
depths in the St. Marys River, Michigan, 1983. Dashed
lines represent productivity on dead stems, solid lines
on live stems (McNabb, 1984).

spawning substrates resulted in resorption of gonads or mortality of eggs due to detachment (Makhotin, 1977).

The major conclusion regarding the effects of water levels on fish communities in Great Lakes wetlands is fairly clear: present data are too fragmentary for understanding all the complexities of how fish communities interact with changing water levels. The level of refinement of data important to isolate and quantify the effects of this single variable is not available. Fish communities interact with many other environmental variables including vegetative patterns, temperature, dissolved oxygen, toxic substances, primary productivity rates, water currents, turbidity, sedimentation, food availability, and others, all of which may be influenced by water levels. Add to this the apparent diversity of fishes in the wetlands, with each species having particular requirements that change with life history stage, and the complexities continue to mount. An apparent important factor, however, is for water levels to increase and remain high during spawning and post-spawning periods of major marsh species (spring/summer).

A beginning research need would be to have a research team at each "typical" wetland (Figure 5) to compile the basics on fish communities at each site throughout the same year(s). Basic questions such as (1) what is the seasonal fish community comprised of (i.e., we know practically nothing about winter use of these habitats); (2) which species spawn there; (3) what is the length of time that post-larval use shallow zones as nursery sites; (4) what are the food webs for both young and adult fish communities; (5) what are the migratory patterns of fish within the marsh; (6) what is the relative proportion (seasonally) of "typical" lake fishes compared to "typical" marsh species in these habitats. Unfortunately, this basic research is often difficult to sell to funding agencies, although it seems unwise to attempt to understand functional aspects and relationships to environmental variables without these kinds of data. Along with the fisheries team a limnological team is needed to collect pertinent physical-chemical data, and information on primary and secondary production. With more basic information in hand, perhaps controlled experiments using manipulated water levels could be developed. Marshes that may lend themselves to such manipulation would be the existing man-made types and possibly the natural "restricted riverine" and "connected lake" types. Literature reviews on effects of water level changes in reservoirs by Hildebrand et al. (1980) and Ploskey (1983) provide good coverage outside the Great Lakes Basin and should be consulted in future water level research in the Great Lakes.

Approximately 18,000 kilometers of Great Lakes shoreline including some 6,600 kilometers of island shoreline exist (IGLLB, 1973). Each kilometer undoubtedly offers habitat and is used by certain species. Our rather primitive and fragmentary knowledge of the importance of these habitats to fishes should be recognized by research/funding agencies and management agencies, and future Great Lakes fisheries research should focus much more on these habitats than it has in the past.

DISCUSSION

William W. Taylor
Department of Fisheries and Wildlife
Michigan State University
East Lansing, Michigan 48824

The paper presented by Dr. Liston and Ms. Chubb is an excellent overview of the relationship between water level fluctuation and habitat quality and quantity. Additionally, it shows the potential for artificially manipulating, where possible, fish population abundance and productivity. Natural changes in water level fluctuations in Great Lakes coastal marshes are generally not as abrupt as those in a reservoir but still impact fish productivity by altering habitat quantity and quality.

This paper represents a reversal, to some degree, by many fisheries ecologists regarding the underlying mechanisms which control fish population dynamics. Clearly, Dr. Liston and his colleague are indicating that environmental factors, in this case the hydrological cycle and morphological characteristics of the particular wetland, are key factors in determining the species and the biomass of each species that can live in a marsh. In my opinion, fisheries scientists and ecologists in general have ignored or minimized the effect of environmental parameters over the past 25 years, instead opting for biotic control mechanisms such as predation and competition. I am not saying that biotic control is not important; I am just agreeing with the veiwpoint that environmental factors can, and often do, have a dramatic impact on fish productivity. In this case, the environmental factor of the hydrologic cycle determines the amount and quality of habitat available for the different species of fish associated with wetlands.

It is important to examine two interrelationships when discussing fish-wetland interactions; fish to the wetlands and wetlands to the fish. Wetlands are highly productive systems and serve as an energy and nutrient sink from the outwelling of river systems. Estuaries and their associated wetlands are noted nurturing areas for many anadromous fish and often serve a vital link in determining year class strength and future yields of these fish (Northcote, 1978). The abundance of aquatic macrophytes in wetlands provides excellent cover and feeding habitat for larval and juvenile fishes. Additional benefits of the marsh to fish are in providing spawning habitat and, in some cases, winter habitat for fish in a river system which are attempting to find more constant and favorable conditions than in their summering habitats.

How does a fish influence a marsh? There is not a lot of information available on this but several possibilities can be proposed. For anadromous fish which die after spawning in river systems, nutrients from their decaying corpses would flow downstream to the river mouths and their associated wetlands (Richey et al., 1975). It has been proposed that these nutrients are important in the spring bloom of phytoplankton and zooplankton which, in turn, provide the food for their out-migrating young of the year. In addition, these nutrients would be important for macrophyte production which provides cover and wetable substrates for food production later in the season.

Fish may also be involved in plant succession/degradation. A clear example of this would be the grass carp (Ctenopharyngodon idella) which feeds on macrophytes and can greatly alter the species present in a wetland (Sutton, 1977). In fact, the grass carp may influence waterfowl productivity as its preferred food tends to be the same as that of the mallard (Anas platyrhynchos) (Martin and Uhler, 1939; Swanson et al., 1979). By influencing the marsh structure, its function is also influenced. This influence is felt not only within the aquatic community but in the waterfowl community. The type of habitat available influences not only fish but wildlife productivity, and the abundance of fish influences the species composition and abundance of waterfowl sustained by a wetland.

Much research still needs to be done on fish-wetland relationships, and I commend Dr. Liston, Ms. Chubb and other researchers at this symposium for undertaking the very difficult task of sampling and deciphering the dynamics of wetland ecosystems. I think if one recognizes that fish are attempting through natural selection to optimize their growth, survival, and reproduction and that these in turn are regulated by both environmental and biotic factors; we will make progress in our understanding of fish-wetland interactions. What fish use the wetland is determined by the niche made available in a wetland at any particular time which, in turn, determines the degree of transitory use and the abundance and productivity of the fish.

LITERATURE CITED

Beam, J.H. 1983. The effect of annual water level management on population trends of white crappie in Elk City Reservoir, Kansas. No. Am. J. Fish. Mgmt. 3:34-40.

Braun, E. 1967. The survival of fish larvae in reference to their feeding behavior and food supply. pp. 113-131 in S.D. Gerking (ed.), The Biological Basis of Fish Production. John Wiley and Sons, New York, NY.

Brazo, D.C. and C.R. Liston. 1979. The effects of five years of operation of the Ludington Pumped Storage Power Plant on the fishery resources of Lake Michigan (1972-1977). Mich. State Univ., Dept. Fish. Wildl., 1977 Ann. Rep. to Consumers Power Co., Vol. II, No. 1. 406 pp.

Brazo, D.C. 1984. Department of Fisheries and Wildlife, Michigan State Univ., East Lansing, MI, personal communication.

Busch, W.D.N. and L.M. Lewis. 1982. Some Great Lakes wetland responses to water level variations and the use of wetlands by fish. Minutes of the Lake Ontario Committee 1982 Annual Meeting, Appendix VII. Great Lakes Fishery Commission.

Carbine, W.F. 1943. Egg production of the northern pike, Esox lucius L., and the percentage survival of eggs and young on the spawning grounds. Mich. Acad. Sci. Arts and Letters. 20:123-137.

Chevalier, J.R. 1977. Changes in walleye (Stizostedion vitreum vitreum) populations in Rainy Lake and factors in abundance, 1924-75. J. Fish. Res. Board Can. 34:1696-1702.

Chubb, S.L. 1984. Spatial and temporal distribution and abundance of larval fishes in a coastal wetland on Lake Michigan. M.S. Thesis, Mich. State Univ., Dept. Fish. Wildl., 185 pp. + Appendix.

Cushing, D.H. 1974. The possible density-dependence of larval mortality and adult mortality in fishes. pp. 103-111 in J.H.S. Blaxter (ed.), The Early Life History of Fish. Springer-Verlag, New York, NY.

Forney, J.L. 1968. Production of young northern pike in a regulated marsh. N.Y. Fish Game J. 15(2):143-154.

Great Lakes Basin Commission. 1975. Levels and flows. Appendix II. Great Lakes Basin Framework Study, Ann Arbor, MI. 206 pp.

Great Lakes Commission. 1984. Great Lakes News Letter, Vol. XXIV, No. 6.

Hassler, T.J. 1970. Environmental influences on early development and year class strength of northern pike in Lakes Oahe and Sharpe, South Dakota. Trans. Am. Fish. Soc. 99:369-380.

Herndendorf, C.H., S.M. Harley, and M.D. Barnes, eds. 1981. Fish and Wildlife Resources of the Great Lakes Coastal Wetlands within the United States. Volume One: Overview. U.S. Fish and Wildlife Service, Washington, DC. FWS/OBS-81/02-V1.

Hildebrand, S.G., R.R. Turner, L.D. Wright, A.T. Szluka, B. Tschantz, and S. Tam. 1980. Analysis of environmental issues related to small-scale hydroelectric development, III: Water level fluctuation. ORNL/TM-7453. Oak Ridge National Laboratory, Oak Ridge, TN. 132 pp.

Hough, J.L. 1958. Geology of the Great Lakes. University of Illinois Press, Urbana. 313 pp.

Hulsey, A.H. 1958. A proposal for the management of reservoirs for fisheries. Proc. Ann. Conf. Southeast Assoc. Game Fish Comm. 12:132-143.

International Great Lakes Levels Board. 1973. Regulation of Great Lakes water levels. Appendix D: Fish, Wildlife and Recreation. Report to the International Joint Commission.

International Lake Erie Regulation Study Board. 1981. Lake Erie Regulation Study: Report to the International Joint Commission.

Jaworski, E., C.N. Raphael, P.J. Mansfield, and B.B. Williamson. 1979. Impact of Great Lakes water level fluctuations on coastal wetlands. U.S.D.I. Office of Water Resources and Technology. 351 pp.

Johnson, D.J. and M.D. Barnes. 1983. Personal communication.

Jude, D.J., F.J. Tesar, J.A. Dorr III, T.J. Miller, P.J. Rago, and D.J. Stewart. 1975. Inshore Lake Michigan fish populations near the Donald C. Cook Nuclear Power Plant, 1973. Spec. Rep. No. 52, Great Lakes Res. Div., Univ. Mich., Ann Arbor, MI. 267 pp.

Jude, D.J., G.R. Heufelder, N.A. Auer, H.T. Tin, S.A. Klinger, P.J. Schneeberger, C.P. Madenjian, T.L. Rutecki, and G.G. Godun. 1980. Adult, juvenile and larval fish in the vicinity of the J.H. Campbell Power Plant, eastern Lake Michigan, 1978. Spec. Rep. No. 79. Great Lakes Res. Div., Univ. Mich., Ann Arbor, MI. 607 pp.

Keith, W.E. 1975. Management by water level manipulation. Pages 489-497. in H. Clepper, (ed.), Black Bass Biology and Management. Sport Fishing Inst., Washington, DC.

Lantz, K.E., J.T. Davis, J.S. Hughes, and H.E. Schafer, Jr. 1964. Water level fluctuation - its effect on vegetation control and fish population management. Proc. Ann. Conf. Southeast Assoc. Game Fish Comm. 18:483-494.

Liston, C.R., D.C. Brazo, R. O'Neal, J. Bohr, G. Peterson, and R. Ligman. 1981. Assessment of larval, juvenile and adult fish entrainment losses at the Ludington Pumped Storage Power Plant on Lake Michigan. Mich. State Univ., Dept. Fish. Wildl., 1980 Ann. Rep. to Consumers Power Co., Vol 1. 274 pp. + appendices.

Liston, C.R. and S.L. Chubb. 1983. Abundance, distribution and ecological relationships of larval and juvenile fishes in the Pentwater Marsh on Lake Michigan. Progress report to Michigan Sea Grant, U.S. Dept. Commerce (Project No. R/CW-13). 30 pp.

Liston, C.R., C. McNabb, D. Brazo, J. Craig, W. Duffy, G. Fleisher, G. Knoecklein, F. Koehler, R. Ligman, R. O'Neal, M. Siami, and P. Roettger. 1984. Environmental baseline studies of the St. Marys River during 1982 and 1983 prior to proposed extension of the navigation season. Mich. State. Univ., Dept. Fish. Wild. Rep. to U.S. Fish Wildl. Serv. 764 pp. + appendices.

Liston, C.R., C. McNabb, W. Duffy, D. Ashton, R. Ligman, F. Koehler, J. Bohr, G. Fleischer, J. Schuette, and R. Yanusz. 1983. Environmental baseline studies of the St. Mary's River near Neebish Island, Michigan, prior to proposed extension of the navigation season. Mich. State Univ., Dept. Fish. Wildl. Rep. to U.S. Fish Wildl. Service, FWS/OBS-80/62.2. 316 pp.

Makhotin, Y.M. 1977. The spawning efficiency of fishes in Kuybishev Reservoir and factors determining it. J. Ichthyol. 17(1):24-35.

Manny, B.A. 1984. Potential impacts of water diversions on fishery resources in the Great Lakes. Fisheries. 9(5):19-23.

Martin, A.C. and F.M. Uhler. 1939. Food of game ducks in the United States and Canada. U.S. Dept. Agric. Tech. Bull. 634, 156 pp.

Mathis, W.P. 1966. Observations on control of vegetation in Lake Catherine using Israeli carp and a fall and winter drawdown. Proc. 19th Ann. Conf. Southeast Assoc. Game Fish Comm. 19:197-205.

McCrimmon, H.R. 1968. Carp in Canada. Bull. No. 165, Fish Res. Board Can., Ottawa, Canada. 93 pp.

McNabb, C. 1984. Aquatic plants and primary production. In C. Liston et al., (eds.), Environmental Baseline Studies of the St. Marys River During 1982 and 1983 Prior to Proposed Extension of the Navigation Season. Mich. State Univ., Dept. Fish. Wildl. Rep. to U.S. Fish Wildl. Service. 764 pp. + appendices.

Miranda, L.E., W.L. Shelton, and T.D. Bryce. 1984. Effects of water level manipulation on abundance, mortality, and growth of young-of-the-year largemouth bass in West Point Reservoir, Alabama - Georgia. N. Am. J. Fish. Mgmt. 4:314-320.

Nelson, W.R. 1978. Implications of water management in Lake Oahe for the spawning success of coolwater fishes. pp. 154-158 in R.L. Kendall (ed.), Selected Coolwater Fishes of North America. Am. Fish. Soc. Spec. Publ. No. 11.

Northcote, T.G. 1978. Migratory strategies and production of freshwater fishes. In: S.D. Gerking (ed.), Ecology of Freshwater Fish Production, John Wiley & Sons, NY, pp. 326-359.

Ploskey, G.R. 1983. A review of the effects of water level changes on reservoir fisheries and recommendations for improved management. Tech. Rep. E-83-3, prepared by the U.S. Fish and Wildlife Service, National Reservoir Research Program, for the U.S. Army Engineer Waterways Experiment Station, C.E., Vicksburg, MS. 83 pp.

Richey, J.E., M.A. Perkins, and C.R. Goldman. 1975. Effects of kokanee salmon (Oncorhynchus nerka) decomposition on the ecology of a subalpine stream. J. Fish. Res. Board Can. 32:817-820.

Ringler, N.H., R.G. Werner, M.S. Kruse, and S.R. LaPan. 1984. Population ecology of northern pike and muskellunge in the St. Lawrence River. Paper presented at the American Fisheries Society 114th Annual Meeting, Ithaca, NY.

Shields, J.T. 1958. Experimental control of carp reproduction through water drawdown in Ford Randall Reservoir, South Dakota. Trans. Am. Fish. Soc. 87:22-33.

Sutton, D.L. 1977. Grass carp (Ctenopharyngodon idella) in North America. Aquatic Botany 3:157-164.

Swanson, G.A., G.L. Krapu, and J.R. Serie. 1979. Foods of laying female dabbling ducks on the breeding grounds. In: T.A. Bookhout (ed.), Waterfowl and Wetlands - An Integrated Review, La Crosse Printing Co., Inc., LaCrosse, WI, pp. 47-57.

Tarzwell, C.M. 1941. Fish populations in the backwaters of Wheeler Reservoir and suggestions for their management. Trans. Amer. Fish Soc. 41:201-214.

U.S. Fish and Wildlife Service. 1982. Lake Ontario Shoreline Protection Study (Sage Creek Marsh). Prepared for the U.S. Army Corps of Engineers, Buffalo, NY. 184 pp.

Westlake, D.F. 1966. Some basic data for investigations of the productivity of aquatic macrophytes. In C.R. Goldman (ed.), Primary Productivity in Aquatic Environments. Mem. Ist. Ital. Hdrobiol., 18 Suppl., University of California Press, Berkeley, CA.

Whillans, T.H. 1979. Historic transformation of fish communities in three Great Lakes bays. J. Great Lakes Res. 5(2):195-215.

CHAPTER 9

SIMPLIFIED COMPUTATION OF WETLAND VEGETATION CYCLES

Robert H. Kadlec and David E. Hammer
The University of Michigan
Wetland Ecosystem Research Group
Department of Chemical Engineering
H. H. Dow Building
Ann Arbor, Michigan 48109

ABSTRACT

Based on data from the Houghton Lake Porter Ranch Wetland, an accounting of biomass, nitrogen and phosphorus is presented, for the natural stationary repetitive state. The budgets for the wetland are constructed from data on ten compartments: annual and woody live biomass, roots, standing dead, annual and woody litter, three soil layers and surface water. A simple set of empirical rules for biomass behavior provide a reasonable description of seasonal variations. A simple computer program allows the calculation of annual cycles, based on material supplies and constraints, and the most commonly measured variables.

RATIONALE

The description of wetland ecosystem vegetation requires accounting for biomass and nutrients, principally nitrogen and phosphorus. The determination of even a small subset of the necessary field data demands considerable effort, and has resulted in fragmented knowledge. Investigations are commonly made separately for litterfall and decomposition, productivity by species, mortality, and other detailed studies. Enough of this type of work has now been done to consider ways of combining and coordinating subsystem results. The ultimate purpose of such an integration of knowledge is to understand how the entire system reacts to human intervention. But it is first necessary to understand the internal allocations of nutrients and the amounts of biomass as a function of time for a stable ecosystem. This is most convenient if available in the form of nutrient and biomass budgets, or material balances.

The number of chemical components (usually at least 3), plus the number of ecosystem compartments (usually at least 4), plus the necessity for following time histories, combines to dictate a computer numerical procedure. The sparsity of data means ad hoc rules and equations which must then be tested in non-direct ways. In return, one must not expect too much detail. The wetland under consideration here is the Porter Ranch peatland, located in Roscommon County, Michigan. It has been the subject of field studies since 1971, with the goal of determining the response to nutrient additions. Current studies are in large part concerned with the effects of the addition of treated wastewater to the wetland (Kadlec 1985a).

MODEL DEVELOPMENT

Compartments and Transfers

The active biomass zone encompasses the aboveground live and dead plant parts, plus the belowground parts, together with the soil in which they are located. There are subsets of this material which are conveniently identifiable and separable in the field. One such breakdown is shown in Figure 1. While not unique, this arrangement fits available data for the peatland under consideration.

Wood biomass has a longer life span than leaves or annual plants. thus it is reasonable to separate live material on this basis, giving rise to live woody and live annual aboveground compartments. The entire annual crop transfers to an annual litter pool, perhaps with losses due to leaching and partial decomposition. In the case of strong stemmed annuals, such as Typha spp., this transfer could be through an annual standing dead compartment. The choice has been made here to identify all dead annual material as litter, and to use the corresponding definition of annual litter decay. The standing annual material does not generally persist for more than one year. In contrast, dead wood stems persist for several years, and in this peatland, for decades. Consequently a standing dead compartment is established for accounting purposes, which includes only woody stems. Eventually, dead wood does fall, forming a woody litter compartment.

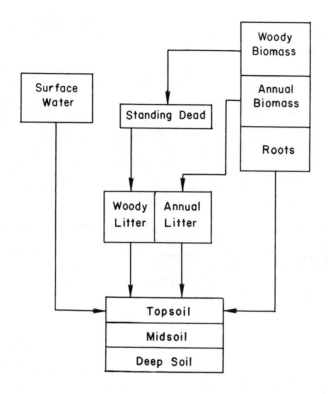

Figure 1. Compartment model of the wetland.

Annual litter and woody litter are depleted by decomposition, but at different rates; the annual being the faster. Some residual fraction of each is incorporated into new soil. Roots increase during the growing season, and decrease due to death of portions of the root mass. Since it is virtually impossible to distinguish dead roots from the rest of the newly formed peat, they are transferred to the soil compartment.

If the peatland is accreting soil, there are several gradual processes which must occur. As the soil column builds, new roots will invade the upper strata, and deep roots will succumb. Thus, on a time scale of years, the soil column builds and the root zone follows. The assumption made here is that a layer of topsoil exists which contains few roots, and which receives sediment and litter decomposition residuals and decomposes at a lesser rate than litter. If this is regarded as a zone of fixed thickness, it then translates vertically upward very slowly. There is no clear distinction between this zone and the litter above, or the midsoil root zone below.

With respect to this slowly moving reference frame, there is an export from topsoil to midsoil of the net accretion in the top zone. This input is augmented by the accumulation of dead root material, and decremented by decompositon. For a fixed midsoil zone thickness, there is therefore an export to deep soil, which supports no further activity. Although this ultimate burial of material is small in terms of vertical movement--typically a few millimeters per year--it is the only significant permanent sink for carbon, nutrients and other materials in the global wetland inputs.

The compartments above also interact with the atmosphere and with surface and interstitial waters to obtain the necessary carbon and nutrients. The assumption here is that the wetland will export excesses, and that year-to-year variability is small--a stationary state except for an increasing surface elevation.

Equations

Much is known about some of the processes described above; so much that it is tempting to incorporate all the detail and subtleties. For example, plant growth can be related to insolation, temperature, translocation, dark respiration, carbon dioxide concentration, leaf resistance; and many other factors. For example, Dixon (1974) constructed a four compartment model, including aboveground vascular plants, standing dead, litter and one soil zone. His 47 equations (several were ordinary differential equations) contained about 80 parameters and required several time function specifications, such as temperature, insolation and atmospheric carbon dioxide levels.

It is possible, but not easy, to estimate all these parameters and functions, and to solve the differential and algebraic equations over a period of several years. Dixon's results are reproduced in Figure 2, for one particular set of parameters. Several important observations may be made concerning these biomass histories. First, it takes 3-7 years of simulation time to "wash out" the assumed initial conditions. The amount of computation could be cut by an order of magnitude if only the final, repetitive state were calculated for one year. Second, the initial conditions, chosen on January 1, were not all close to the stationary state values on January 1, as for litter,

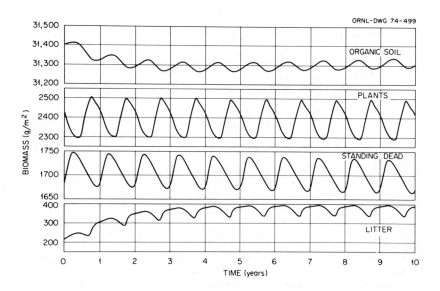

Figure 2.   Ten year simulation of peatland biomass (from Dixon, 1974).

for example. That is not surprising, because there are no data on litter biomass on that date for any northern wetland, and it was therefore, of necessity, guessed.

Third, the smooth computed time functions could be replaced by connected segments of simple curves, such as straight lines or parabolas, with very little loss of detail. There are no data reported for any northern wetland that are good enough to refute such elementary curve shapes. Finally, there are a small number of dates each year on which the curves change form, corresponding to the seasonal patterns of growth, litterfall and decomposition.

It is possible to preserve most of the desired detail while vastly simplifying the model. The connection times between such elementary, calculated segments are chosen to be the start of the growing season, the time of peak annual standing crop, and the end of annual litterfall. Thus each compartment is constrained to at most three periods of different behavior: spring-summer, fall, and winter.

A balance equation may be written for each compartment:

Bushes $\dot{B} = G_{B} - D_{B}$ (1)

Annuals $\quad \dot{A} = G_{A} - DA$ $\qquad$ (2)

Roots $\quad\quad \dot{R} = G_{R} - DR$ $\qquad$ (3)

Standing Dead $\dot{S} = D_{B} - Ds$ $\qquad$ (4)

Annual Litter $\dot{L} = D_{A} - DL$ $\qquad$ (5)

Woody Litter $\dot{W} = D_{S} - Dw$ $\qquad$ (6)

Peat $\quad\quad \dot{P} = R_{S+fRDR+fLDL+fWDW-DMSPMS-DTSPTS}$ $\qquad$ (7)

This is the same starting point as more complex models, but the similarity ends here. The rates of growth and death are assumed to be simple _time_ functions, with parameters determined from _easily obtained_ field data. For most compartments and time periods, nothing more than straight lines are warranted. Thus, one set of choices for G and D functions are sets of constants over the three time periods. The number of such constants is 44; consisting of 36 values of G,D,R; 3 values of f; 3 times; and P_{MS} and P_{TS}. Hence, there is still a need for considerable information.

PARAMETER EVALUATION

This is a complex situation, with many choices for the use of data, hence the following should be viewed as a site-specific example.

In addition to field data, there are constraints which assist in determination of parameters. The stationary state requires that all compartments except deep soil return to their starting biomass at the end of one year. This is equivalent to six parameter specifications. Growth of live biomass does not occur to any appreciable extent in winter, or in some cases in fall either, yielding zero rates. It is not easy to distinguish between fall and winter rates of decomposition, but winter rates are much slower. In recognition of these seasonal effects, winter and fall rates of decomposition are considered equal, and the summer rate is a fixed multiple of the winter rate, α. Sparse information on root biomass indicates some need for simplicity in the corresponding model. A fixed ratio of root to shoot growth was presumed in the present study for both annual (a) and woody (b) material.

The peak standing crops of the biomass compartments can be determined more easily than rate parameters, and

Table 1. An example of parameter requirements. Time is zero on January 1 and is measured in years. There are 29 pieces of information.

Component	Spring–Summer	Fall	Winter
Biomass			
Woody growth	G_B	0	0
Woody death	D_B	D_B	D_B
Annual growth	G_A	0	0
Annual death	0	D_A	0
Root growth	$aG_A + bG_B$	0	0
Root death	D_R	D_R	D_R
Annual litter decomposition	αD_L	D_L	D_L
Standing dead fall	D_S	D_S	D_S
Woody litter decomposition	αD_W	D_W	D_W
Soil			
Sedimentation	R_S	0	0
Midsoil decomposition	αD_{MS}	D_{MS}	D_{MS}
Topsoil decomposition	$\beta \alpha D_{MS}$	βD_{MS}	βD_{MS}
Nonseasonal			
Times:	t_1, t_2, t_3		
Soil:	P_{TS}, P_{MS}; F_R, F_L, F_W		
Initial values:	B_o, A_o, R_o, L_o, W_o, S_o, P_o		

were utilized to provide those six pieces of information about the annual cycles. The annual biomass is zero at the beginning of spring, which completes the biomass specification as shown in the top of Table 1.

Table 2.   Parameter determination from field data.   There are 29 specified values.

Indirect

Periodicity (6): $B(t+1) = B(t)$,...etc.

Peak values (6): $A(t_2) = A_{peak}$,...etc.

Direct

Initial values (2): $P(o)=o$, $A(o)=o$

Final values (1): $P(1)=P_e$

Biomass parameters (4): a, b, α, D_B

Soil parameters (7): $P_{TS}, P_{MS}; f_R, f_L, f_W; \beta, R_S$

Times (3): t_1, t_2, t_3

     The soil compartments are less well understood. The inventories in topsoil and midsoil are chosen to be constant, and may be specified on a mass or volume basis. The residual fractions from litter decay must be known together with sedimentation rates.   There is likely a lessening of soil decomposition with depth, leading to a prescribed ratio between topsoil and midsoil (β).   It is most likely that data will be available on total accretion rate, manifested by the deep soil year-end accumulation leading to the ability to compute a decomposition rate, as indicated in Table 1.

     The net result of this data organization is the ability to calculate the amount of biomass in each compartment as a function of time in a stable, stationary state. For the example in Table 2, Figure 3 shows the qualitative appearance of the resulting piecewise linear approximation to the biomass behavior.

     There is little difficulty in relaxing the conditions of linear behavior.   In this study, there was sufficient data to warrant a growth curve for annual biomass during the spring-summer season.   The linear assumption of constant growth yields a linear equation for annual biomass in spring-summer:

$$G_A = \frac{A_{PEAK}}{t_2 - t_1} \tag{8}$$

$$A = A_{PEAK} \frac{t - t_1}{t_2 - t_1} \tag{9}$$

The field data are better fit by a square root relation:

$$G_A = \frac{A_{PEAK}}{\sqrt{2}\ t_2 - t_1} \frac{1}{\sqrt{t - t_1}} \tag{10}$$

$$A = A_{PEAK} \frac{t - t_1}{t_2 - t_1} \tag{11}$$

Such improvements need require no further information in the form of new parameters.

JUSTIFICATION

Some of the above assumptions appear quite restrictive, but several can be supported by site data. Aboveground live leatherleaf (Chamaedaphne calyculata) was found to increase linearly from May to September (Richardson, et al., 1976), but bog birch (Betula pumila) displayed more erratic behavior. This later behavior may have been due to inadequate sampling, since Wentz (1976) found similar high variability for willow (Salix spp.) and attributed it to the insufficiency of the one m^2 plots. More recent data on stem diameters shows a constant rate of increase throughout the summer for willow and bog birch (Kadlec, 1985a). The winter behavior of aboveground wood has not been studied, but overall mortality must compensate for growth in a stationary state.

Evidence for the existence of a repetitive state is shown in Figure 4, which shows the annual growth curves for sedge (Carex spp.) measured in different years. The Wentz (1976) data are not significantly different from the Kadlec (1985a) data 10-11 years later at this site. Further, the square root growth curve for spring-summer and linear death relation for fall are supported for this annual. The beginning of the growing season is clearly close to May 1, the peak crop occurs close to September 1, and all live annual biomass is gone by November 1. Changes of ± 15 days in these times cause only ± 5 percent changes in calculated annual biomass.

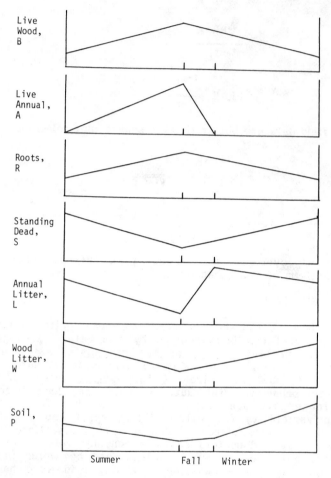

Figure 3.   Qualitative annual cycles, represented by line segments.

Belowground biomass for leatherleaf was found to be 28±7 percent of the total, independent of season (Wentz and Chamie, 1980).   After the spring low annual biomass period for sedge (early June), and before the fall, low biomass period (mid-September), Wentz (1976) found the sedge root/shoot ratio to be 7.4±1.2 on seven dates. In the absence of further information, it seems reasonable to proportion above- and below-ground growth rates.   An obvious improvement could and should be made if data became available for translocations.

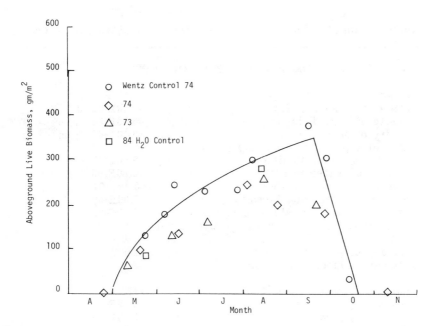

Figure 4. Live above-ground sedge.

Annual litter biomass follows the pattern of Figure 3 (Chamie, 1976), although leatherleaf, willow and bog birch show a gradual late summer to fall increase in leaf-fall. Sedges tend to die over a much briefer period, forming an annual standing dead pool which is here lumped with annual litter. The litter decomposition processes have been studied for wood and leaf litter and support the assumption of slow winter rates (Chamie, 1976; Kadlec, 1985a).

Reliable information on standing dead wood is not available due to sampling difficulties, but it comprises about 10-15 percent of live wood based on data.

Soil accretion has been determined by carbon-14 and cesium-137 testing methods (Kadlec, 1985b). The rates are on the order of a few millimeters per year.

IMPLEMENTATION AND USE

This set of equations was solved using simple algorithms on a digital computer (Hammer, 1984). No integrations are necessary, but there is a considerable amount of algebraic

Table 3. Input information for the stationary state simulator.

THE FOLLOWING DATA WERE USED FOR INPUT

Thickness of topsoil layer	=0.10 E+00 m
Thickness of midsoil layer	=0.10 E+00 m
Density of topsoil layer	=0.10 E+04 kg/m^3
Density of midsoil layer	=0.10 E+04 kg/m^3
Ratio of top to midsoil decay rates	=0.10 E+02
Fraction of water in litter	=0.90 E+00
Fraction of water in soil	=0.80 E+00
Ratio of root to ann. biomass growth rate	=0.10 E+01
Ratio of root to wdy. biomass growth rate	=0.30 E+00
Ratio of summer to winter decay rate	=0.10 E+02
Exponent for annual biomass growth model	=0.50 E+00
First day of growing season	=120
Last day of growing season	=258
Last day of annual litter fall	=300
Yearday for initial conditions	=120
Peak annual biomass	=0.35 E+00 kg/m^2
Peak woody biomass	=0.15 E+00 kg/m^2
Initial standing dead	=0.23 E−01 kg/m^2
Peak annual litter	=0.11 E+01 kg/m^2
Peak woody litter	=0.87 E+00 kg/m^2
Frac. of roots decaying to soil solids	=0.25 E+00
Frac. of litter decaying to soil solids	=0.20 E+00
Death rate for woody biomass	=0.27 E−03 kg/m^2/d
Initial annual biomass to root ratio	=0.90 E+00
Initial woody biomass to root ratio	=0.50 E+00
Sedimentation deposition rate	=0.0 kg/m^2/d
Net soil accumulation rate	=−.80 E−05 kg/m^2/d

---

computation and management of input and output. This program returns the fundamental rates from the data supplied, so that comparisons and judgments may be made on alternative bases. Calculations can then be made of the biomass in any compartment on any yearday. Sample input is shown in Table 3, output in Tables 4 and 5.

A significant use of this simulation is to determine the nutrient allocations in the ecosystem. Additional data is entered for the nitrogen and phosphorus concentrations in each compartment, and thus the standing crop of each may be computed by simple multiplication. Any deficits must be supplied from external sources or internal storages in the interstitial and surface waters. Inputs with precipitation are computed from data on amounts and concentrations.

Table 4. Biomass versus time for the Porter Ranch peatland, stationary natural state calculations.

Year-day	Annual Biomass	Woody Biomass	Root Biomass	Annual Litter	Woody Litter
120	0.0	0.88E-01	0.64E+00	0.10E+01	0.87E+00
130	0.94E-01	0.15E+00	0.71E+00	0.10E+01	0.87E+00
140	0.13E+00	0.15E+00	0.73E+00	0.98E+00	0.87E+00
150	0.16E+00	0.15E+00	0.74E+00	0.95E+00	0.87E+00
160	0.19E+00	0.15E+00	0.74E+00	0.93E+00	0.87E+00
170	0.21E+00	0.15E+00	0.74E+00	0.91E+00	0.87E+00
180	0.23E+00	0.15E+00	0.74E+00	0.89E+00	0.87E+00
190	0.25E+00	0.15E+00	0.74E+00	0.87E+00	0.87E+00
200	0.27E+00	0.15E+00	0.73E+00	0.85E+00	0.87E+00
210	0.28E+00	0.15E+00	0.73E+00	0.82E+00	0.87E+00
220	0.30E+00	0.15E+00	0.72E+00	0.80E+00	0.86E+00
230	0.31E+00	0.15E+00	0.71E+00	0.78E+00	0.86E+00
240	0.33E+00	0.15E+00	0.70E+00	0.76E+00	0.86E+00
250	0.34E+00	0.15E+00	0.70E+00	0.74E+00	0.86E+00
260	0.33E+00	0.15E+00	0.69E+00	0.74E+00	0.86E+00
270	0.25E+00	0.15E+00	0.69E+00	0.82E+00	0.86E+00
280	0.17E+00	0.14E+00	0.68E+00	0.90E+00	0.86E+00
290	0.83E-01	0.14E+00	0.68E+00	0.98E+00	0.86E+00
300	0.0	0.14E+00	0.68E+00	0.11E+01	0.86E+00
310	0.0	0.14E+00	0.68E+00	0.11E+01	0.86E+00

Internal storages occur during the winter, when decomposition products are immobilized beneath the ice. These may be recovered, at least in part, before being flushed out of the ecosystem.

All of the above processes need not create balanced situations for nutrients. For nitrogen, there remain questions concerning fixation or denitrification, involving atmospheric absorption and release. However, phosphorus is more constrained. The model parameters may not be chosen in a manner which creates unreasonable imports or exports, based on further knowledge of the surface water phenomena. Consequently, the summary sections of the program provide the necessary calculations for balancing nutrient allocations. In the example shown, there is slight excess of both N and P available for export.

Also available from such computations is the time history of each nutrient pool, thus providing the "nutrient cycle" information. Temporary shortages and excesses are

Table 5.   Nutrient distributions calculated for the stationary state.

UNPERTURBED STATIONARY-STATE ANALYSIS:

Availability of Nitrogen During the Growing Season:

Requirement for growth = 0.8954E-02 kg/m^2

Input with rain	=	0.1481E-03 kg/m^2	(1.65%)
Litter recovery, annual	=	0.2404E-02 kg/m^2	(26.85%)
Litter recovery, woody	=	0.6441E-04 kg/m^2	(0.72%)
Roots decay recovery	=	0.1750E-02 kg/m^2	(19.55%)
Soil release/decay	= ·	0.3518E-02 kg/m^2	(39.29%)
Surface water exchange	=	0.0 kg/m^2	(0.0 %)
Shortage(-)/excess(+)	=	-0.1068E-02 kg/m^2	(-11.93%)

Winter recovery inputs	=	0.1400E-02 kg/m^2	(15.63%)
Annual budget closure	=	0.3312E-03 kg/m^2	(3.70%)
Net Precip. less evap.	=	0.7370E-01 m	

Availability of Phosphorus During the Growing Season:

Requirement for growth = 0.6576E-03 kg/m^2

Input with rain	=	0.7404E-05 kg/m^2	(1.13%)
Litter recovery, annual	=	0.2404E-03 kg/m^2	(36.56%)
Litter recovery, woody	=	0.1804E-04 kg/m^2	(2.74%)
Roots decay recovery	=	0.1979E-03 kg/m^2	(30.09%)
Soil release/decay	=	0.1104E-03 kg/m^2	(16.78%)
Surface water exchange	=	0.0 kg/m^2	(0.0 %)
Shortage(-)/excess(+)	=	-0.8347E-04 kg/m^2	(-12.69%)

Winter recovery inputs	=	0.9957E-04 kg/m^2	(15.14%)
Annual budget closure	=	0.1609E-04 kg/m^2	(2.45%)
Net Precip. less evap.	=	0.7370E-01 m	

---

reasons to re-examine model parameters, since shortages alter the growth pattern and excesses may be exported via outflow. It should be noted that even a crude seven compartment model has the tendency to create nearly constant total inventories of biomass, nitrogen and phosphorus.

Large scale simulations of ecosystem response to external driving forces require a starting condition. Even the simplest simulation requires very large amounts of CPU time (money) to calculate for a short span of time (Parker, 1974). If initial conditions are chosen which are not consistent with a reasonable starting annual cycle, then several years of simulation are needed simply to recover from the bad starting guesses. To avoid this difficulty, it is possible to compute a final stationary state for a periodic disturbance, such as a repeated addition of nutrient-rich treated wastewater (Gupta, 1977). If the external driving forces are not periodic--and they rarely are--then the simulation must start at a viable repetitive starting state. This model provides such a capability.

CONCLUDING REMARKS

The model and computer program described here have proven absolutely necessary as the point of departure for calculating the response of a wetland system to external upsets in flow and nutrients. But coincidentally, there is a capability for doing biomass and nutrient budgets at frequent intervals without tedious hand calculations. Considerations of nutrient import and export can be used in conjunction with most easily available field data to estimate the intrinsic rates.

The complexity of ecosystem simulation requires that something be sacrificed to achieve these ends. This model abandons cause and effect submodels for growth, death and decay processes. In so doing, there can be no response to intraseasonal environmental factors. Further, there must exist a stable and repetitive set of driving forces, such as nutrient concentrations, photoperiods, temperatures, and the like. It is possible to use this same framework to consider such effects, but simulation then becomes aperiodic.

LITERATURE CITED

Chamie, J.P.M. 1975. The Effects of Simulated Sewage Effluent Upon Decomposition, Nutrient Status and Litterfall in a Central Michigan Peatland. Ph.D. Thesis. University of Michigan, Ann Arbor. 110 pp.

Dixon, K.R. 1974. A Model for Predicting the Effects of Sewage Effluent on a Wetland Ecosystem. Ph.D. Thesis. University of Michigan, Ann Arbor. 111 pp.

Gupta, P.K. 1984. Dynamic Optimization Applied to Systems with Periodic Disturbances. Ph.D. Thesis. University of Michigan, Ann Arbor. 225 pp.

Hammer, D.E. 1984. An Engineering Model of Wetland/Wastewater Interactions. Ph.D. Thesis. University of Michigan, Ann Arbor. 139 pp.

Kadlec, R.H. 1985a. Wetland Utilization for Management of Community Wastewater, 1984 Operations Summary, Houghton Lake, MI. Report to HLSA and MDNR, 100 pp.

Kadlec, R.H. 1985b. Sediment Processes in Wetlands. Submitted to Water Research.

Parker, P.E. 1974. A Dynamic Ecosystem Simulator. Ph.D. Thesis. University of Michigan, Ann Arbor. 305 pp.

Richardson, C.J., J.A. Kadlec, W.A. Wentz, J.P.M. Chamie, and R.H. Kadlec. 1976. Background ecology and the effects of nutrient additions on a central Michigan wetland. In: Proceeding of the Third Wetland Conference, M.W. LeFor, W.C. Kennard, and T.B. Helfgott, (eds.), Institute of Water Resources, University of Connecticut, Report No. 26, pp. 34-72.

Wentz, W.A. 1975. The Effects of Simulated Sewage Effluents on the Growth and Productivity of Peatland Plants. Ph.D. Thesis. University of Michigan, Ann Arbor. 112 pp.

Wentz, W.A. and J.P.M. Chamie. 1980. Determining the belowground productivity of Chamaedaphne calyculata, a peatland shrub." Int. J. Ecol. Environ. Sci. 6:1-4.

NOTATION

a	annual root/shoot ratio
A	annual biomass, kg/m2
b	woody root/shoot ratio
B	woody biomass, kg/m2
D	death or decay rate, yr-1
f	residual fraction
G	growth rate, yr-1
L	annual litter biomass, kg/m2
P	peat biomass, kg/m2
R	root biomass, kg/m2
S	standing dead biomass, kg/m2
t	time, years
W	woody litter biomass, kg/m2

GREEK

α	ratio of summer to winter decay rates
β	ratio of topsoil to midsoil decay rates

SUBSCRIPTS

A	annual
B	bush or woody shrub
e	end of year
L	annual litter
MS	midsoil
PEAK	peak
R	roots
S	standing dead
TS	topsoil
W	woody litter
1	start of growing season
2	start of leaf fall
3	end of leaf fall

CHAPTER 10

WETLAND VALUATION: POLICY VERSUS PERCEPTIONS

Patricia B. Weber
Eastern Michigan University
Ypsilanti, Michigan

INTRODUCTION

Traditional wetland valuation strategies have been based upon financial models expanded to frame such resource economics issues as valuing the imputed costs of environmental policy alternatives. These cost-benefit analyses utilize present value techniques to examine discounted cash flows, payback periods or profitability indices as a method to establish the comparative advantage of land use alternatives. Finance-based models are credible evaluation tools for investment alternatives which possess identifiable cash flows or streams of benefit. However, their applicability to land use problems which require estimation of social value rather than private values is less than complete because of at least two shortcomings: (1) traditional financial models offer no provision for the measurement or estimation of affective, nonmonetary values attached to alternative uses; and (2) the comparison of benefit streams or returns on investment are estimates of the variable costs and returns to the parcel in use and do not reflect the land owner's perceptions of the worth of a parcel (as distinct from its market value).

It is common knowledge that property holders may invest disproportionate sums in a parcel relative to their expected returns on that investment. Indeed, an earlier Michigan Sea Grant study on consumer investment in shoreline protection (Braden and Rideout, 1980) indicated that landowners were willing, in some cases, to invest beyond the total cash or market value of their property.

In the face of behavior which violates traditional investment standards, financial models provide little guidance as policy makers structure incentives to motivate landowners to favor environmentally sound land use practices.

The attempts of federal and state agencies to establish a socially optimal balance of wetlands throughout the Upper Midwest region is but one case in point. The recent work of Leitch (1983) on the prairie pothole region of the Upper Midwest and the prairie provinces of Canada represents a classical, and particularly insightful cost-benefit analysis in a financial management framework. Concentrating upon a seven-county area in south and west central Minnesota, Leitch interviewed farm operators who had drained wetlands during recent years. Interviews were conducted between February and March, 1981. Separate area studies were conducted to sort the cost differential by drainage technique, i.e., random wetland drainage versus subsurface tile drainage (n=35 and n=62, respectively).

Cash estimates of private landowner returns to wetland drainage included increased crop sales, decreased nuisance or avoidance costs and a component for the net influence of intangibles.

Increased crop sales were estimated using a present value algorithm based upon discounted cash flow. Extensive computations were based on variable costs of production (not return on land values). Using traditional cost-benefit analysis techniques, Leitch found that many wetland drainage activities affecting agricultural land in Minnesota were economically feasible within boundaries of the 8–12 percent interest rate examined.

The cost-benefit analysis fails to account, literally, for owned worth of the land, in addition to the potential net benefits or costs of alternative uses, and therefore risks stimulating interest in wetland drainage. Leitch observed that, even in the face of monetary incentives to preserve wetlands which were in excess of the net present value of draining the wetlands, many farmers declined U.S. Fish and Wildlife Service and U.S. Department of Agriculture incentive payments in favor of draining owned wetlands (Leitch, 1983). Leitch speculated that other nonmonetary costs or benefits were influencing farmers' decisions to drain agricultural wetlands. Yet the methodolgy employed was unable to identify, quantify, or prioritize such nonmonetary considerations as instrinsic worth of a parcel, leaving policy decision makers with little guidance in program formulation to encourage wetland preservation.

One alternative, of course, would be to simply increase monetary incentives gradually until participation was optimal. However, such tactics tend to elicit counterstrategies on the part of landowners who may attempt to estimate "peak" payoffs and drive the incentives payment higher in an artificial market. It would be far preferable strategically, and in terms of total social cost, to asses an adequate cash value for nonmonetary considerations and shift the incentive structure in a one-time adjustment, rather than to invoke a bidding posture.

NONMONETARY VALUES ASSESSMENT

In an attempt to gain an understanding of the prioritization and quantification of traditionally nonmonetary values confronted in wetlands preservation, a study[1] utilizing perceptual scaling techniques was developed for application to a Michigan constituency of wetlands activists, managers and ecologists on the one hand, and real estate developers, farmers, news media and political constituents on the other. Ten intensive interviews of representatives of each constituency were designed to intersperse qualitative and quantitative measures. Respondents were asked to estimate both the array and intensity of values attached to traditionally nonmonetary and nonquantifiable attributes of wetlands in the context of a decision to preserve or develop wetland parcels.

RESEARCH METHODOLOGY

An exploratory study of this nature is enhanced by the extent to which it can describe variations in perceptions held by conflicting constituencies, for parcels of widely divergent characterization, and for which agreed-upon attributes can be assigned and measured. The procedure used in the larger study called a joint-space technique, applies to a small-sample metric spatial analysis to first characterize and array wetland types (see Figure 1).

---

[1]Coastal Wetlands in Conflict: A Joint-Space Approach for Assessing the Perceived Utility of Wetlands, Patricia B. Weber (Ann Arbor, Michigan Sea Grant Program), 1981. This publication is a result of work sponsored by the Michigan Sea Grant Program, project number R/CW-1, with a grant, NA-80-AA-D-00072, from the Office of Sea Grant, National Oceanic and Atmospheric Administration (NOAA), U.S. Department of Commerce, and funds from the State of Michigan.

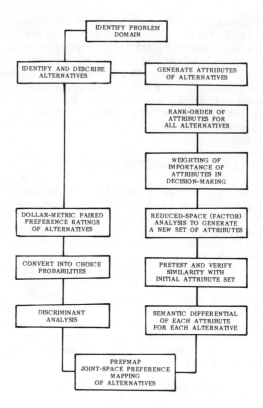

Figure 1. Analytical procedure for preference-mapping social alternatives.

Next, a preference-mapping function weighs subjective values assigned to wetland types by expert judges who represent each constituency. This mapping schema delivers a spatial analysis which superimposes the attitudes of expert judges on the array of wetlands and quantifies values assigned according to a distance algorithm.

The purpose of this paper is somewhat narrower. Our concern is to assess the intrinsic worth of wetland parcels in an exercise which accounts for nonmonetary attributes. Therefore, we will focus upon the original data collection of both wetland attributes and upon the results of dollar metric evaluations of perceived worth.

Analysis Plan

The initial task in defining the domain of the problem was to identify and describe wetland types considered to be endangered in Michigan. Subsequently, the relevant constituencies were identified and asked to provide a qualified spokesperson to serve as an "expert judge" of wetland types. (While a more broadly based survey could sample representative opinion from the constituency at large, in political environments, it is often useful to compare opinions of leaders who negotiate legislative and regulatory content.) Last, but not least, a list of relevant attributes of wetlands was solicited to provide dimensions for comparing and contrasting wetland types. With the parameters of the research problem defined, each expert judge would be asked to participate in an in-depth personal interview to examine a range of parcels serving as candidates for preservation or development. The personal interviews were structured to obtain both cognitive (evaluative) and affective (preference) perceptions of each parcel.

Wetland Types

Prior to field data collection, it was necessary to formulate a more detailed description of alternative types of wetlands prevalent in Michigan and other Great Lakes states. Certainly, since the study was focused upon identifying jeopardized classes of wetlands and attaching a degree of subjective value to each, priority was placed upon classification schemes which emphasized endangered wetlands rather than all wetland types.

Two wetland researchers and two public officials engaged in resource management were asked to provide classification schemes. Although original classifications varied somewhat, after second interviews in a Delphi procedure all reached a consensus that the following six wetland types were both present in Michigan and, to greater or lesser degrees, endangered by developmental pressures:

(1) Agricultural wetlands—embedded on parcels otherwise committed primarily to agricultural use.

(2) Great Lakes coastal wetlands—wetlands contiguous with the Great Lakes and its primary channels, including rivermouth wetlands, flood plains, etc.

(3)   Wetlands on an inland lake--wetlands contiguous with a large expanse of open water found inland of coastal waterways.

(4)   Riverbank wetlands--wetlands contiguous with river channels, such as riverine, flood plain and swamp.

(5)   Forested or shrub-covered wetlands--wetlands chiefly characterized by woody vegetation which have no major contiguous waterway (e.g., cedar swamps).

(6)   Peat or moss-covered bogs--usually isolated wetlands covered by organic masses with unstable surfaces.

The wetland categories are presented in the approximate order in which they are found in the environment in Michigan. Most prevalent are agricultural wetlands, least prevalent are peat or moss-covered bogs. The listing presents no judgment of the intrinsic or extrinsic value of any of the wetland types, since that is the task to be asked of study participants.

Constituencies

The sample frame consisted of those interest groups with broad, statewide concerns rather than groups with a limited, local perspective. As a general guide for the selection of constituencies to be included in the sample frame, those groups actively lobbying for (and against) Michigan Public Act 203 (1979)--the "wetlands" bill--were included.

According to House and Senate committee members (and their legislative aides), the following groups were most visible in support of Public Act 203 (1979): (1) East Michigan Environmental Action Council, (2) West Michigan Environmental Action Council, (3) Sierra Club, (4) the Audubon Societies, and (5) Michigan Farm Bureau.

The East and West Michigan Environmental Action Councils were the most active representatives of the preservationists' philosophies. Well-organized and with a long record of political action at both state and local levels, both Environmental Action Councils are effective lobby organizations. While they share the preservationist philosophy, the Audubon Societies (both Detroit Audubon Society and the Michigan Audubon Society) have more specific organizational objectives than those of the Environmental Action Councils.

The Audubon Societies are comprised primarily of members who are recreational birders. They are non-resident day users of primary wetland areas who are devoted to passive observation rather than more active forms of recreation.

The Sierra Club, unlike its local preservationist counterparts, brings to bear a nation-wide constituency. While it shares many of the concerns of the Action Councils, it is more likely to represent the interest of individual, rather than institutional members.

Although the Michigan Farm Bureau was identified as a supporter of Public Act 203 (1979), it is true that wetlands in largely rural areas were exempted from permit requirements as a result of earlier amendments to the bill. It is not clear whether the agricultural constituency would remain in favor of a more extensive wetland management program, should one be proposed.

Those constituencies expressing opposition to Public Act 203 (1979) were virtually all representatives of residential, commercial and industrial development interests: (1) Michigan Realtors Association, (2) Michigan Home Builders Association and (3) Michigan Iron Mining Association.

The real estate professions and association developers share an interest in increasing both the quality and intensity of land use. The most organized constituency representing this group is the Michigan Realtors Association, whose members are real estate brokers (some of whom are also land developers). Residential developers are also represented by the Michigan Home Builders Association, a group of contractors engaged primarily in subdividing undeveloped land for both speculative and custom building. The Michigan Iron Mining Association (a trade association representing mining companies concentrated in the Upper Peninsula) is an example of industrial concerns expressed during consideration of the bill.

The League of Women Voters and the print media, two public interest constituencies, joined the organizations active in following the legislative progress of the bill. The League of Women Voters is a voter information organization committed to full and complete disclosure of knowledge to the voting public. While expressly nonpartisan, the organization does publicize a stance on proposals presented before the electorate or in the legislature. The print media in general provides coverage of legislative activities in both news and editorial sections. While both print and audio-visual media participated to a greater or lesser extent in the coverage of the "wetlands bill," one metropolitan daily newspaper was mentioned most frequently as a source of information and opinion on the issue.

Although the organizations identified are not the only ones which took a public position on wetland management, they are the groups which represented key constituencies identified by a social network analysis of the debate surrounding passage of the wetlands bill, as were the "expert judges" selected from each constituency. The inclusion of these organizations represents the dispersion of opinion toward preservation/development held by public interest groups, environmental activists, wetland users, agricultural land users, and residential, commercial and industrial developers.

Sample Selection

In a research problem such as this where individuals represent the opinions and attitudes of a generalized constituency, it is important that "expert Judges" be selected carefully. A social network analysis of interactions was conducted as a part of participant selection to identify those individuals most influential within their organization or constituency when wetland management issues are addressed. It was felt that careful identification of participants and opinion leaders in the legislative process would enhance the validity of the results and overcome possible weaknesses in the application of the expert judgment technique.

WETLAND ATTRIBUTES

To arrive at a set of commonly understood attributes which might influence the decision to preserve or develop a particular parcel, each of the sixty-three participants in the supporting social network study was asked to describe important functions or characteristics of wetland parcels. The participants were interviewed by telephone with comprehensive notes supporting their conversations. Any attribute or characteristic mentioned was transcribed to note cards and sorted by topic. Redundant attributes were deleted and similar features merged to yield a master list of 23 wetland attributes (see Table 1).

Twenty-five members of the original social network analysis were called a second time and asked to rate the importance of each characteristic for deciding whether to preserve or develop a wetland parcel. The attributes were rated on a scale ranging from "1 = Not at all important" to "5 = Extremely important." The ratings were subject to principal components analysis to reduce the original

Table 1. Master list of wetland attributes.

---

1. Size
2. Proximity to other bodies of water
3. Proximity to an urban area
4. Recharges the water supply (traps runoff)
5. Purifier of water (natural filtration system)
6. Water reservoir (supports water table)
7. Density and diversity of species
8. Habitat for wildlife, birds, fish, wetland plants
9. Breeding ground for wildlife, birds, fish
10. Recreation value in present state
11. Recreation value when developed
12. Beauty as unique scenic areas or greenbelt
13. Importance for tourism (improves attractiveness of Great Lakes and inland lakes)
14. Remains a wetland year-round
15. Historically (over many years) a wetland
16. Uniqueness of a given wetland
17. Fragileness of a given wetland
18. Great expense of developing wetland
19. Potential public health nuisance (mosquitoes)
20. Cost of replacing natural wetland functions
21. Financial potential of land
22. Possibility of complementary uses
23. Natural flood control mechanisms

---

set of 23 attributes to a number more manageable for judgments collected in round-robin comparisons. That is, the original set of attributes would have required n(n-D/2) or 253 separate comparisons taken without regard to the order of presentation. Previous field research had indicated that more than 19 attributes were likely to fatigue the respondent and generate a resistance to the remainder of the interviewer's schedule (Braden, 1974 and 1980).

The results of principal components analysis yielded 18 components which extracted 99.42 percent of the variance in the original set of 23 attributes. All of the components accounted for significant variance reduction at $\alpha = 0.05$. The correlations between original attributes and derived components were used to redefine a reduced set of 18 attributes to use in further data collection exercises.

The interpretation of the principal components of the wetland preservation is of value apart from later application in field activities. Table 2 details the

Table 2. Derived attributes of wetland preservation decision;
Principal components and signifance of variance extracted.

---

1. Habitat and breeding ground for wildlife (0.0000)
2. Proximity to other bodies of water (0.0000)
3. Recreation value and beauty when developed (0.0000)
4. Extent to which parcel remains a wetland year-round (0.0000)
5. Financial potential when developed (0.0000)
6. Potential public health nuisance (mosquitoes) (0.0000)
7. Expense of developing wetland (0.0000)
8. Usefulness as a recreation attraction (tourism) (0.0001)
9. Density and diversity of species supported (0.0002)
10. Extent to which other uses interfere with wildlife and habitat (fragileness) (0.0010)
11. Ability to support multiple, complementary uses (0.0030)
12. Number of years parcel has been a wetland (0.0046)
13. Value as an urban greenbelt or scenic area (0.0085)
14. Size (0.0100)
15. Usefulness as a water reservoir (supports water table) (0.0132)
16. Usefulness for flood control (0.0135)
17. Usefulness for providing water supply (traps runoff, recharges aquifers) (0.0181)
18. Recreation value in its present state (0.0252)

---

reduced set of wetland attributes and provides the associated
significance of each to variance reduction. The first
component, "Habitat and breeding ground for wildlife,"
accounted for nearly one-third of the variance (32.18%)
observed among the original wetland attributes. For those
participants in the wetlands controversy in Michigan,
this factor was of primary importance in determining whether
a wetland should be preserved or developed. The remaining
factors can be interpreted to appear in decreasing order
of their importance in the preservation/development decision
according to the consensus of this diverse group.

Clearly, the passive functions of wetlands as water
reservoirs or for flood control or water supply were rated
as less intrinsic to the preservation decision than active
social uses for either the human or wildlife populations.
While this suggests a difference in the perceived value
of passive wetland functions, the results also reflect
the intangible nature of passive wetland attributes.
As such, the passive functions may simply have less direct
and more diffuse an influence on the perceptions of individuals

Table 3. Attributes affecting preservation decision.

ENVIRONMENTAL CONSTITUENTS

1. Providing habitat and breeding ground for wildlife
   (98.67 MAS and 39.41 EMEAC)
2. Relative cost of developing a wetland
   (72.26 WMEAC)
3. Potential financial returns as a recreation site
   (56.40 DAS)
4. Ability to support multiple uses
   (44.01 SC and 42.31 EMEAC)
5. Passive or water-related wetlands benefits
   (38.17 SC)
6. Relative size of the parcel
   (36.86 DAS)

DEVELOPMENT CONSTITUENCIES

1. Potential for recreational purposes
   (99.93 MFB)
2. Use as a habitat and breeding ground for wildlife
   (97.16 MAR)
3. Lack of "nuisance wetness"
   (63.88 MHBA)

PUBLIC REPRESENTATION

1. Ability to sustain multiple uses
   (78.96 PM)
2. Resilience of wetland in use
   (52.43 LWV)
3. Extent of water cover
   (33.23 LWV)

---

than those characteristics which affect participation in active use of a parcel. That the listing reflects an ordering of self-interest rather than community interest may reflect participants' assessments of the availability of wetland services from these and alternative or substitute providers.

Each constituency reflected its own compelling rationale for preservation. In those groups distinguished by their environmental activism, a combination of tangible costs and benefits derived from alternative use served to support justifications derived from more traditional wetlands functions (Table 3). Development constituencies offered grounds for negotiation by considering habitat contributions, although they found recreation potentials an equally compelling argument. A nuisance component was the only other factor which figured strongly in the development constituencies' decision. Public representatives were noteworthy for the strong focus on long-range social welfare which was reflected in their decision based upon wetland resilience and capacity.

RESULTS OF SUBJECTIVE VALUATION

When approaching a preservation or development decision, any individual or group must ultimately be prepared to exercise its preferences in order to gain control over the outcome of a particular parcel. Although this research problem addresses generic wetland types rather than particular parcels, respondents were asked to relate how much they perceived they would have to pay (in current dollars per hectare) to purchase a parcel predominantly of each wetland type. Participants were pressed to be as realistic as possible within the limits of their knowledge. The estimates tendered were treated as subjective valuations of the cost of gaining control of the parcel for preservation or development.

It is noteworthy that the relative dollar amounts reported by each subject reflect that subject's perception of the value placed upon each wetland type by the external, competitive market environment. They are discrete estimates of the dollar metric equivalent required to convert that parcel to the respondent's own use. Differences among respondents' estimates for the same generic wetland type include variability in market knowledge and risk aversion, which are reflected across all estimates rendered by a given subject, as well as differences resulting from variations in the subjective perception of the worth of a particular classification of wetlands.

Although respondents were asked to provide a best estimate of market value, the data are perceptual in nature. They are not only influenced by the market knowledge of each respondent but by their biases toward alternative land use. In addition, each estimate was presented by an "unwilling seller." That is, market estimates generated by an individual who does not intend to sell (as is the case with many wetland holders) are likely to be inflated. The difference between transactions value and perceived value is a quantitative estimate of intangible values accruing to ownership. The perceived value is assumed to approximate the maximum cost per acre of redeeming wetlands in each category.

Table 4. Perceived values of generic wetland types (Michigan
Sea Grant Wetlands Study, 1981).

	Perceived Value1,2 ($/ha)	Percent of Average Perceived Value	Standardized Value Across Subjects3
Agricultural Wetland	$2,170	(54.4)	0.509
Great Lakes Coastal Wetland	9,778	105.6	1.862
Wetland on an Inland Lake	7,563	59.0	1.437
Riverbank Wetland	5,921	24.5	1.235
Forested or Shrub-covered Wetland	1,995	(58.0)	0.591
Peat or Moss-covered Bog	1,109	(76.7)	0.364
Average Perceived Value	$4,755		1.000

1Based on mean perceived price per hectare in response
to the question: "In your opinion, what is the average
price per hectare that you would have to pay today for
a parcel that is predominantly (generic wetland type)?"

2Significant differences in perceived values observed
across wetland types: $F_{(2,27)} = 12.252$,   = 0.01).

3Significant differences in standardized values observed
across wetland types: $F_{(2,27)} = 149.960$,   = 0.01).

Values which were quoted ranged from an estimate
of $185 per hectare for a peat or moss-covered bog to
$37,037 per hectare for Great Lakes coastal wetlands.
Table 4 provides the mean perceived values reported for
each wetland type. Clearly, respondents perceived that
some wetlands were significantly more valuable than others.
Besides the two extremes of bogs and coastal wetlands,
it would appear that wetlands which are contiguous with
other bodies of water are perceived to be of greater worth
than wetlands which are "typically" isolated upland.
The average dollar value attached to wetlands generally
was $4,755 per hectare.

Because subjects can place widely divergent values
on the same generic wetland type, estimates of subjective

value were standardized across respondents to remove the
effects of interpersonal variation in dollar metrics.
Table 4 illustrates that only the ordering of agricultural
wetlands and forested or shrub-covered wetlands is affected
by controlling for respondent variation. The standarized
value of wetland types reflects the indexed average relative
value of wetlands. With individual variations in the
dollar metric removed, agricultural wetlands dropped behind
forested or shrub-covered wetlands in value, although
the dollar value increased to $2,421 ($4,755 x 0.509).
The differences in average perceived value reflect significant
differences in the relative value attached to generic
wetland types for both standardized and nonstandardized
scales.

DISCUSSION AND RECOMMENDATION

The problem addressed in this study is of potential
significance to resource managers and the public at large.
The ability to decompose the likelihood of preserving
or developing generic wetland types, as well as specific
wetland parcels, into cognitive and affective components
of utility may affect deployment of both financial and
professional resources in support of either outcome.

In the problem presented for analysis, participating
constituencies placed greatest priority upon gaining control
of Great Lakes coastal wetlands, followed at some distance
by wetlands on inland lakes and riverine wetlands. The
effort the participants were willing to expend to influence
the use of coastal wetlands far exceeded that for any
other generic wetland type. That effort was also a priority
among competing constituencies and forewarned of heated
financial, political and/or legal confrontation to gain
control of particular parcels of mutual interest. In
such an arena, extramural interest groups and the news
media may play key roles in building the total force of
public opinion in favor of a particular outcome.

The plight of agricultural wetlands, forested or
shrub-covered wetlands and peat or moss-covered bogs is
clear as physical scientists unfold the nature of their
contributions as habitat. Although physical evidence
supports the need for preserving agricultural wetlands
for migratory flyways and for conserving the unique habitats
of cedar swamps and peat bogs, the lack of broad public
interest in support of a preservation effort leaves such
wetlands vulnerable to alteration or development.

The large difference in perceived value across wetland
types suggests that policy makers should distinguish preservation

priorities and establish an incentive structure which reflects the perceived value of differing types of wetlands. The U.S. Fish and Wildlife Service incentive payment of $847 per hectare or approximately $242 per acre for preserving agricultural wetlands in the prairie pothole region from drainage and crop use compares poorly with the perceived value of $2,171 per hectare ($2,420 per hectare, standardized) for Michigan agricultural wetlands.

Such an incentive payment must also account for the potential income foregone by the landowner. In the case of the farmers in the prairie pothole region, agricultural wetlands may generate additional net benefits, the present value of which may range from $780 to $3,842 per hectare at 12 percent interest over a 15-year life (Leitch, 1983). Clearly, the $847 per hectare incentive provides little motivation to restrain drainage. In the case of Great Lakes coastal wetlands, the cost to the public of retiring prime coastline from development may require substantial financial commitment to cover both the perceived worth of a parcel and projected cash flow.

It is reassuring that the public does modify judgment regarding the priority for preservation by the potential of a wetland for active human and animal use. To improve the probability of preserving critical wetlands, resource managers might emphasize practices which intensify active rather than passive functions of wetlands, so long as passive benefits are not foregone.

LITERATURE CITED

Abdalla, C.W. and L.W. Libby. 1982. Economics of Michigan Wetlands. Agricultural Economics Report No. 410, Department of Agricultural Economics, Michigan State University, East Lansing, MI.

Braden (Weber), P.L. 1975. Integrating the interests of competing groups in socio-economic analyses: A suggested approach. Proc. of the Ninth Annual Conference of the Marine Technology Society, Washington, DC.

Braden (Weber), P.L. 1980. Cost-benefit pricing of new products using joint-space scaling techniques. Proc. of the Business and Economics Statistics Section, American Statistical Association, Houston, TX.

Braden (Weber), P.L. 1974. Joint-space resolution of public policy problems: A case study in nuclear energy. Proc. of the Business and Economics Statistics Section, American Statistical Association, Houston, TX.

Braden (Weber), P.L. and S.R. Rideout. 1980. Consumer Investment in Shore Protection. Michigan Sea Grant Program, Ann Arbor, MI.

Brande, J. 1980. Worthless, valuable or what? An appraisal of wetlands. J. Soil and Water Conserv. 35(1):12-16.

Jones, W.H. 1973. An overview of the Krannert 1972-73 interviewing program on social awareness. Prepared as a classroom exercise at the Krannert School of Industrial Administration, Purdue University, West Lafayette, IN.

Jones, W.H. 1973. Developing discriminant configurations from small samples. Paper 408, Institute for Research in the Behavioral, Economic and Management Sciences, Krannert Graduate School of Industrial Administration, Purdue University, West Lafayette, IN.

Jones, W.H. 1974. Single subject discriminant configurations: An examination of reliability, validity, and joint-space implications. Paper 451, Institute for Research in the Behavioral, Economic and Management Sciences, Krannert Graduate School of Industrial Administration, Purdue University, West Lafayette, IN. Presented at the Psychometric Society Meetings, Stanford University.

Kreisman, J., J. McDonald, G. Rosenbaum, and J. Snyder. 1976. Coastal Wetlands: With Emphasis on Freshwater Systems. The University of Michigan School of Natural Resources Regional Planning Program, Ann Arbor, MI.

Leitch, J.A. 1983. Ecomonics of prairie wetland drainage. Trans. of the American Society of Agricultural Engineers.

Michigan Department of Natural Resources. Undated. Wetlands and You. (Pamphlet).

CHAPTER 11

ONTARIO'S WETLAND EVALUATION SYSTEM
WITH REFERENCE TO SOME
GREAT LAKES COASTAL WETLANDS

Eleanor Bottomley
Canadian Wildlife Service
Ottawa, Ontario

ABSTRACT

"An Evaluation System for Wetlands of Ontario South
of the Precambrian Shield" was produced jointly by Environment
Canada and the Ontario Ministry of Natural Resources (EC/OMNR,
1983). The evaluation system is designed to numerically
quantify wetland values to permit comparison of wetlands
relative to each other. The evaluation system is broad
in perspective: it can be applied to four wetland types--marshes,
swamps, fens and bogs--and it encompasses four categories
of wetland values--biological, social, hydrological and
special features.

Vigorous field testing and statistical analysis
of evaluation results showed that the system is reproducible,
and it appears to produce a fairly accurate ranking of
wetlands. Marshes, swamps and bogs scored fairly highly
for the Biological Component; Social Component scores
covered a range of values for all wetland types; Hydrological
Component scores for marshes were consistently low whereas
bogs obtained much higher scores; and Special Features
Component scores were high for many wetlands including
a large number of marshes.

Wetlands are grouped into seven classes on the basis
of evaluation scores, with Class 1 and 2 wetlands being
the most valuable. Of the 30 Great Lakes coastal wetlands
evaluated on Lakes Ontario, Erie and St. Clair, 19 (63%)
were Class 1 and 2 wetlands, and 90 percent were Class

3, 2 or 1. The high performance of these coastal wetlands derives from their strengths in the Biological and Special Features Components; Hydrological Component scores were very low. Details of the scoring system are illustrated using Second Marsh (Oshawa, Ontario) as an example.

INTRODUCTION

A wetlands evaluation system has been developed for use in southern Ontario which allows one to quantify numerically a diverse array of wetland attributes or values. This report briefly reviews the development and testing of the wetlands evaluation system, compares the performance of some marshes, swamps and bogs evaluated using the system and summarizes the evaluation scores of 30 Great Lakes coastal wetlands.

Wetland values can differ greatly between wetlands, wetlands being very heterogeneous systems. Therefore, there is often a need for a mechanism or framework through which conflicting claims regarding wetland values can be resolved. One solution to such a problem is to use an evaluation system which will numerically quantify wetland values to permit comparison of wetlands relative to each other, thus facilitating knowledgeable land use decisions.

THE EVALUATION SYSTEM

"An Evaluation System for Wetlands of Ontario South of the Precambrian Shield" was produced jointly by Environment Canada and the Ontario Ministry of Natural Resources (EC/OMNR, 1983). Federal involvement in the development of this system stems in part from the mandate of the Canadian Wildlife Service to protect critical migratory bird habitat, recognizing the importance of wetlands along the lower Great Lakes to the maintenance of healthy migratory waterfowl populations. Development of the wetland evaluation system and the Province of Ontario's on-going program in wetlands management have been discussed by Glooschenko (1983; also see Chapter 12, this volume).

The wetland system applies to the area of Ontario south of the Precambrian Shield. Within this area, the system is applied to four types of wetlands--marshes, swamps, fens and bogs. The evaluation system is broad in perspective, considering an array of wetlands values on a provincial scale and pertaining to different sectors of society. Wetland values are grouped into four separate components--biological, social, hydrological and special features. Each component is evaluated individually and separately from the others, and may generate a total of

250 points. Subcomponent values are weighted to reflect their importance relative to each other. The weighting of subcomponents was determined by experience and judgment.

Biological Component

In this component, the wetland is assessed primarily with respect to its function as wildlife habitat. Biological productivity provides a measure of the ability of a certain area to produce a crop of living organisms, and includes such parameters as type(s) of wetland, soils, growing degree-days, and nutrient status of surface waters. The subcomponent diversity assesses factors which should enable a wetland to support more wildlife species, and includes vegetation communities, interspersion of vegetation and ratio of open water to vegetation. The size subcomponent produces a correlation between diversity and size of the wetland, based on the assumption that a larger wetland is more likely to possess valuable features.

Social Component

This component considers some of the economic, recreational and scientific uses of wetlands. The subcomponents cover harvestable (renewable) resources such as timber, wild rice and bait fish; value for recreational activities such as hunting, fishing, boating or nature appreciation; aesthetics (absence of pollution); use of the wetland for education, scientific studies and nature appreciation; proximity to urban areas, which assesses an urban wetland as having more social value than a wetland in a wilderness location; public ownership/accessibility, which holds that more people benefit from positive values if the wetland is publicly owned; and finally a correlation of wetland size with values that are size-dependent.

Hydrological Component

We have heard in other talks, of the lack of firm data on hydrological relations in wetlands. This caused a problem in deriving values for this component. In the end, only three hydrological values were accepted for evaluation. The first of these hydrological values is the stabilization of flow of rivers and streams. This is based on the premise that wetlands act like basins and can accumulate water during floods and release it in various ways over a more extended time. Flow stabilization is primarily a function of headwater wetlands; Great Lakes coastal wetlands are not assessed for flow stabilization. The second hydrological value of wetlands is water quality improvement, which assesses the capability of a wetland to remove nutrients from surface waters in summer as well as to permanently tie up nutrients in sediments. The third hydrological value of wetlands is their contribution to erosion control, which is a function of the wetland vegetation.

Special Features Component

Whereas the Biological Component looked at the wetland as habitat for wildlife, the Special Features Component evaluates the actual presence of certain species in the wetland. Two subcomponents are evaluated. The rarity/scarcity subcomponent evaluates the rarity of the actual wetland type itself in the region, as well as the presence of significant species in the wetland. The second subcomponent, significant features and fish and wildlife habitat, evaluates features that are of exceptional importance in the public mind. For example, wetlands have value as places where large numbers of waterfowl concentrate to feed and rest during migration. They are also important as fish spawning and nursery areas.

DEVELOPING AND TESTING THE SYSTEM

The wetland evaluation system was developed through a number of stages. First, Ecologistics Limited (1981) assembled a number of existing evaluation models, including Golet (1976), Larson (1976), Cowardin et al. (1979), Reid et al. (1980), Thibodeau and Ostro (1981), Canada Land Inventory (1976) and Gupta et al. (1976). To this was added some new information applicable to the wetlands of southern Ontario, which culminated in the printing of a Draft Evaluation System. This draft system was vigorously field tested in 1982 by Environment Canada and Ontario Ministry of Natural Resources personnel and others, with 110 wetlands being evaluated, including 30 replications. Revisions were made to the draft system resulting in the printing of the First Edition (EC/OMNR, 1983) of the wetland evaluation system. Because revisions were extensive, it was necessary to field test the system again in 1983. In all, 48 separate wetlands were evaluated once or more (Figure 1). Testing was designed to determine the reproducibility and accuracy of the system. Reproducibility was tested by independently evaluating 32 wetlands twice, and analyzing the replicated data statistically. It was shown that, while substantial differences among scores occurred, overall these differences were not statistically significant (i.e., the results of one evaluation team were reproducible by another independent team) (Collins and Maltby, 1984). To assess accuracy, we wanted to know if the scoring system was giving a correct ranking of wetlands. Baseline data

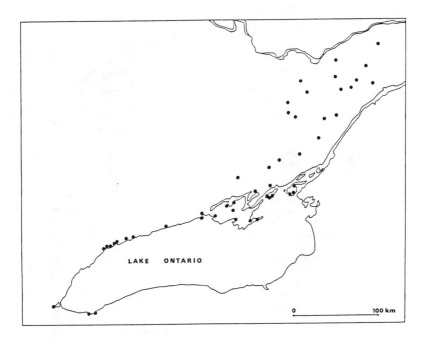

Figure 1. Location of the 48 wetlands evaluated in eastern Ontario and along the Lake Ontario shoreline.

were collected on 35 wetlands including some well-known benchmark wetlands. The sample included swamps and bogs in eastern Ontario and marshes fringing the north shoreline of Lake Ontario. Wetlands were scored and ranked, and indeed a fairly accurate ranking of wetlands was obtained (Maltby et al., 1983).

The distribution of observed scores for marshes, bogs and swamps for each of the four components is shown in Figure 2 (data for 48 wetlands including replications). Biological Component scores were high for all wetland types, and in particular marshes scored highly. Social Component scores for marshes, swamps and bogs covered a range of values indicating that this component is not influenced by wetland type. For the Hydrological Component, there was a large difference in scores between marshes (low) and bogs (high) mainly because of the subcomponent flow stabilization, for which Great Lakes coastal marshes score zero. A range of scores was observed for the Special Features Component. A large number of marshes scored full points (250) for this component, which is an indication

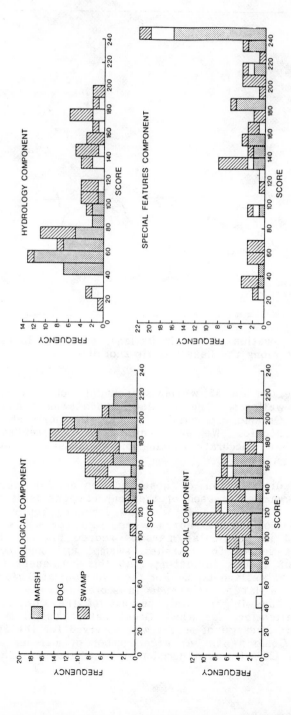

Figure 2. Distribution of observed scores for marshes, bogs, and swamps for each of the four components (from Collins and Maltby, 1984).

Table 1. Ranking wetlands using evaluation scores.

Class	Total Points	Component Scores
1	700 or more	3 components higher than 200 points
2	650 or more	2 components higher than 200 points
3	600 or more	1 component higher than 200 points
4	550 or more	4 components higher than 100 points
5	500 or more	3 components higher than 100 points
6	450 or more	2 components higher than 100 points
7	all others	

(From Ontario Ministry of Natural Resources, 1984. Guidelines for Wetlands Management in Ontario.)

of the high value of these selected marshes as habitat for significant species, and for waterfowl and fisheries.

GREAT LAKES COASTAL WETLANDS

To facilitate the ranking of wetlands, wetlands can be grouped into seven classes on the basis of either total score or component scores. For example, a Class 1 wetland can be obtained either by scoring a total of over 700 points, or by having at least three out of four component scores higher than 200 points each (Table 1). Using these seven classes, we can summarize the evaluation scores for a sample of Great Lakes coastal wetlands.

Thirty wetlands were evaluated, ranging in location from Prince Edward County in the east, along the north shorelines of Lakes Ontario and Erie, to Lake St. Clair (Figure 3). These wetlands ranged in size from a relatively small marsh of 23 hectares to a very large wetland complex of over 3,000 hectares. Nineteen of these 30 wetlands, or 63 percent, were Class 1 and 2, and 27 of the 30 (90%) were Class 3, 2 or 1. From this result, three generalizations can be made. First, coastal wetlands selected for evaluation were mostly large, good quality, benchmark wetlands. Second, in choosing a good quality wetland for evaluation, it will indeed achieve a good ranking using this wetland

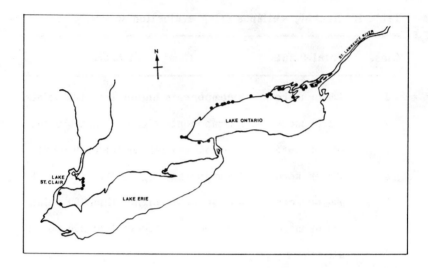

Figure 3.   Location of the 30 marshes evaluated along the Lower Great Lakes shorelines.

evaluation system.   Third, this selection of Great Lakes wetlands is obviously significant in the overall picture of Ontario wetlands (swamps, bogs, fens, marshes) to which this evaluation system applies.

The strengths of these Great Lakes coastal wetlands lay in the Biological and Special Features Components, where mean scores were 200±15 and 236±28 points, respectively

Table 2.   Evaluation statistics for 30 Great Lakes coastal wetlands.

Component	Mean	Standard Deviation	Range
Size (hectares)	354	±608	23 - 3088
Biological Component (points)	200	± 15	145 - 223
Social Component	158	± 32	98 - 214
Hydrological Component	43	± 18	26 - 120
Special Features Component	236	± 28	157 - 250

Table 3. Wetland evaluation scores for Second Marsh.

Component	Maximum Allowable	Observed	Total
BIOLOGICAL			
Productivity	80	71	
Diversity	120	88	
Size	50	50 -----	209
SOCIAL			
Resource Products	60	39	
Recreation	70	20	
Aesthetics	25	10	
Education	35	10	
Proximity	20	20	
Ownership	20	14	
Size	20	18 -----	131
HYDROLOGICAL			
Flow Sabilization	190	0	
Water Quality	35	29	
Erosion Control	25	0 -----	29
SPECIAL FEATURES			
Rarity/Scarcity	250	250	
Significant Features/			
Habitat	250	23	
Ecological Age	15	1 -----	250

(Table 2). Most of these wetlands were large in size,
having a mean area of 354±608 ha. These wetlands were
weak in the Hydrological Component because of their physiographic
position relative to the large surface area of the Great
Lakes.

Table 3 shows scoring details for an evaluation
of Second Marsh, Oshawa, Ontario. The wetland is Class
2 because two components scored higher than 200 points.
The scoring was influenced by the fact that approximately
40 percent of Second Marsh is a central, unvegetated shallow
open water marsh area. This unvegetated area caused fewer
evaluation points to be scored in several components.
Diversity is lower because there is less vegetation interspersion
and a lower ratio of vegetation to open water. Erosion
control is zero because of the absence of vegetation in
the dominant area to buffer erosive forces (40% unvegetated,
31% emergents, 29% trees or shrubs). Also use of the
area by migratory waterfowl has declined, apparently due

to deteriorated habitat conditions. The open area in Second Marsh was created by high Lake Ontario levels in the mid-1970s.

In summary, Ontario's wetland evaluation system is broad in perspective, reproducible and fairly accurate. The system is applicable to Great Lakes coastal wetlands; although hydrology scores are low, coastal wetlands perform well overall because of their strengths in the Biological and Special Features sections.

ACKNOWLEDGMENT

I thank G. McCullough and the Ontario Ministry of Natural Resources for providing wetland evaluation scores for some Great Lakes coastal wetlands.

LITERATURE CITED

Canada Land Inventory. 1976. Land Capability for Agriculture, Preliminary Report No. 10. Environment Canada, Ottawa.

Collins, B. and L. Maltby. 1984. Statistical Analysis of "An Evaluation System for Wetlands of Ontario, First Edition (1983)." Canadian Wildlife Service, unpublished report.

Cowardin, L.M., V. Carter, F.C. Golet, and E.T. LaRoe. 1979. Classification of Wetlands and Deepwater Habitats of the United States. U.S. Fish and Wildlife Service, USDI, Washington, DC.

EC/OMNR. 1983. An evaluation system for wetlands of Ontario south of the Precambrian Shield, first edition. Canada/Ontario Steering Committee on Wetland Evaluation, Environment Canada and Ontario Ministry of Natural Resources. Part I. Rationale, 50 pp. Part II. Procedures Manual 74 pp. Part III. Data Record, 15 pp. Part IV. Evaluation Record, 5 pp.

EC/OMNR. 1984. An evaluation system for wetlands of Ontario south of the Precambrian Shield, second edition. Canada/Ontario Steering Committee on Wetland Evaluation, Environment Canada and Ontario Ministry of Natural Resources. Part I. Rationale and Procedures, 72 pp. Part II. Scoring, 21 pp. Part III. Data Record, 25 pp. Part IV. Evaluation Record, 6 pp.

Ecologistics Limited. 1981. A Wetland Evaluation System for Southern Ontario. Prepared for the Canada/Ontario Steering Committee on Wetland Evaluation and the Canadian

Wildlife Service, Environment Canada.   Includes a Procedures Manual.

Glooschenko, V. 1983.   Development of an Evaluation System for Wetlands in Southern Ontario (presented at Second Annual Meeting of the Society of Wetland Scientists, St. Paul, MN, June, 1983). Wetlands 3:192-200.

Golet, F.G. 1976.   Wildlife wetland evaluation model. In: J.S. Larson (ed.), Models for Evaluation of Freshwater Wetlands. Water Resources Research Centre, University of Massachusetts.

Gupta, T.R. and J.H. Foster. 1976.   Economics of freshwater wetland preservation. In: J.S. Larson (ed.), Models for Evaluation of Freshwater Wetlands. Water Resources Research Centre, University of Massachusetts.

Larson, J.S. (ed.). 1976. Models for Evaluation of Freshwater Wetlands. Water Resources Research Centre, University of Massachusetts.

Maltby, L.S., G.B. McCullough, and E.Z. Bottomley. 1983. Baseline Studies of 35 Selected Southern Ontario Wetlands, 1983. Canadian Wildlife Service, unpublished report.

Reid, R.A., N. Patterson, L. Armour, and A. Champagne. 1980.   A Wetlands Evaluation Model for Southern Ontario. The Federation of Ontario Naturalists.

Thibodeau, F.R. and B.D. Ostro. 1981.   An economic analysis of wetland protection. J. Environ. Manage. 12:19-30.

CHAPTER 12

CHARACTERISTICS OF PROVINCIALLY SIGNIFICANT WETLANDS
AS ASSESSED BY
THE ONTARIO WETLAND EVALUATION SYSTEM

Valanne Glooschenko
Wildlife Branch
Ontario Ministry of Natural Resources
Queen's Park, Toronto, Ontario M7A 1W3

ABSTRACT

Southern Ontario wetland loss is associated with an accompanying decline in wildlife populations. An evaluation system for wetlands in southern Ontario developed by the Ontario Ministry of Natural Resources and the Canadian Wildlife Service, Environment Canada is being used by the provincial government to examine remaining wetlands. Wetlands are ranked by biological, social, hydrological and special features values. By the end of 1984, 700 wetlands had been evaluated across southern Ontario; 94 wetlands were ranked provincially significant (Class 1 and 2) and 84 were regionally significant (Class 3). Ranking of wetlands will be used in guidelines for wetland management.

Characteristics of provincially significant wetlands are discussed by wetland type and physiographic site with reference to their evaluation scores. The hydrological component had considerable influence on the scores of inland swamps and marshes while it contributed little to scores for lakeshore wetlands. The special features component was very important in determining class rank. Important differences in special features subcomponent scores between swamps and marshes were observed; these subcomponents included breeding and feeding by provincially significant animals, winter cover for wildlife, waterfowl staging and fish spawning and rearing.

INTRODUCTION

Some areas of southern Ontario have lost between 81 and 100 percent of their original wetlands from the time of settlement in the mid-1850s to the present (Figure 1). A number of studies in the last several years have commented on this. McCullough (1981) discussed the critical value of the remaining Lake Erie wetlands to migratory waterfowl and Whillans (1979) addressed the fisheries value of lakeshore marshes. Cox (1972), Reid (1979, 1981) and Bardecki (1981) have commented on various aspects of wetland loss. Lynch-Stewart (1983) discussed the greater wetland loss for southwestern Ontario than for the eastern and central portions of the province. Snell (1985) updated and summarized the pressures on southern Ontario wetlands and provided a comprehensive study on wetland distribution and mapping.

A Canada/Ontario Steering Committee for Wetland Evaluation was established in 1981 (D. Euler, Chairperson). An evaluation system for Ontario wetlands south of the Precambrian Shield (EC/OMNR, 1983; 1984) and a provincial wetland class-ranking system have been described by Glooschenko (1983). The basis of the evaluation system is the grouping

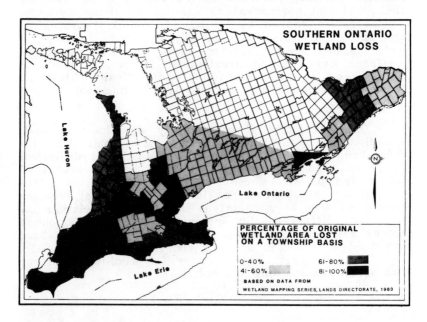

Figure 1. Southern Ontario Wetland Loss

of wetland values according to biological, social, hydrological and special features components. The special features component includes assessment of endangered and provincially significant animals and plants, and habitat for migratory birds. Since completion of the Second Edition of the Evaluation System for Wetlands of Ontario (EC/OMNR, 1984), the Ontario Ministry of Natural Resources has been the lead agency in identifying and evaluating wetlands throughout southern Ontario.

Wetland evaluations were carried out in 1983 by the Ontario Ministry of Natural Resources, the Canadian Wildlife Service and a number of Conservation Authorities, provincial agencies with a mandate for watershed management. In 1984, the Ontario Ministry of Natural Resources and the Conservation Authorities staff continued the wetland evaluation process, and 700 wetlands were evaluated by the end of the year. Quality control in the methodology of wetland evaluation is maintained through evaluation courses coordinated by the Province.

The provincial wetland ranking system provides the basis for Guidelines for Wetland Management issued by the Government of Ontario in April, 1984 (Table 1). Priority was given to the identification and evaluation of the most significant wetlands first. A joint program between the Ontario Ministry of Natural Resources and World Wildlife Fund (Canada) was established in 1984 to summarize information in an interim report on provincially and regionally significant wetlands (Classes 1 to 3) evaluated to date (Glooschenko et al., 1985). Ninety-four of these were ranked as provincially significant and 84 as regionally significant (Table 2). This paper discusses some aspects of scoring of the four components used in evaluation, reasons for high values reached and differences in scoring for some values between wetlands of different dominant type and site for wetlands across southern Ontario.

METHODS

Statistical analyses for this report were carried out on data for 149 Class 1-3 wetlands surveyed by the end of 1984. The Ontario Ministry of Natural Resources and Conservation Authorities staff evaluated 122 wetlands, and 27 were evaluated by the Canadian Wildlife Service. A computerized statistical analysis system (SAS Inc., 1982) was used for the analysis.

In the Wetland Evaluation System (EC/OMNR, 1984) four different wetland types were recognized: marsh, swamp, bog and fen. Single wetlands often contain a combination

Table 1. Wetland classes and scoring.

Class	Total Points		Scoring
Class 1	700 or more	OR	3 out of 4 components scoring higher than 200 points each
Class 2	650 or more	OR	2 out of 4 components scoring higher than 200 points each
Class 3	600 or more	OR	1 out of 4 components scoring higher than 200 points each
Class 4	550 or more	OR	all 4 component scores above 100 points each
Class 5	500 or more	OR	3 out of 4 components scoring more than 100 points each
Class 6	450 or more	OR	2 out of 4 components scoring more than 100 points each
Class 7	all others not included above		

From: Guidelines for Wetland Management in Ontario, April, 1984, Ontario Ministry of Natural Resources.

of wetland types, but for the purpose of this report, a wetland was classified by its dominant type only. Seventy-five of the 149 wetlands surveyed were marshes, 69 were swamps, 3 were bogs and 2 were fens. Because of the small number of bogs and fens surveyed, they were not considered in any analysis dealing with wetland type. In this paper, "marshes" are defined as wet areas permanently or periodically inundated with standing or slowly moving water. They are characterized by robust emergents, and to a lesser extent, anchored floating plants and submergents. "Swamps" are wooded wetlands where standing to gently flowing waters occur seasonally or persist for long periods on the surface. The vegetation cover may consist of coniferous or deciduous trees, tall shrubs, herbs and mosses.

Table 2. Provincially and regionally significant wetlands evaluated in 1983–84.

Region	Number Evaluated	Number Ranked Class*			Percentage Ranked Class 1 & 2 Provincially Significant	Percentage Ranked Class 3 Regionally Significant
		1	2	3		
Central	177	17	17	23	19.2	12.9
Eastern	295	23	22	47	15.2	15.9
South-western	228	8	7	14	6.6	6.1
Subtotal		48	46	84		
Total	700	94		84	13.4	12.0

*Classes 1 and 2 are provincially significant; Class 3 is regionally significant.

In this report, wetlands are also grouped according to their site. Lakeshore wetlands are those on the shore of one of the three Great Lakes in southern Ontario (Lakes Erie, Ontario and Huron). Inland wetlands are hydrologically separated from the influence of the Great Lakes.

Statistical analyses on aspects of the wetland evaluation system scoring by Green (1984) and Collins and Maltby (1984) suggested that the scores for the biological, social and hydrological components approximated normal distributions. However, the special features component scores were quite non-normally distributed. This precludes the possibility of using parametric statistical techniques for analyses involving the special features component.

Analyses of variance and t-tests were used to compare the three approximately normally distributed components among wetlands of different site and/or type. The GT2 and Tukey-Kramer methods (Sokal and Rohlf, 1981) were used for unplanned pairwise comparisons when analyses of variance showed significant differences. For comparing the scores of special features components between wetlands of different type and/or site 2-sample or k-sample median tests were used. Wilcoxon Rank Sums Tests were used to compare the scores of special features subcomponents between marshes and swamps. An alpha level of 0.05 was used for all statistical tests, except for the simultaneous inference methods GT2 and Tukey-Kramer. These methods employ experimentwise error ratio based on the number of unplanned pairwise comparisons which are made.

RESULTS

There was no difference in biological component nor social component scores between marshes and swamps. Hydrological scores were significantly higher for swamps than for marshes (p < 0.001). Special features component scores were significantly higher in marshes than in swamps (p < 0.001).

A significant difference was found between scores for inland and lakeshore sites for the hydrological (p < 0.001) and special features components (p = 0.029). There was no difference in the scores for the biological component nor the social component.

Figures 2 and 3 illustrate comparison of both wetland site and type. Physiographic location of wetlands, whether inland or lakeshore, strongly influenced scoring outcome. Analysis of variance of the three approximately normally distributed components (biological, social and hydrological values) showed significant differences between inland marshes, lakeshore marshes, inland swamps, and lakeshore swamps only for the hydrological component. Pairwise comparisons showed significant differences between inland

Mean Score

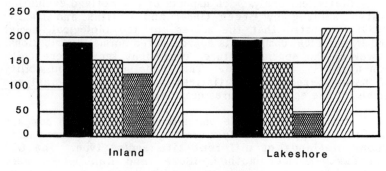

Figure 2. Southern Ontario swamp evaluation scores.

Mean Score

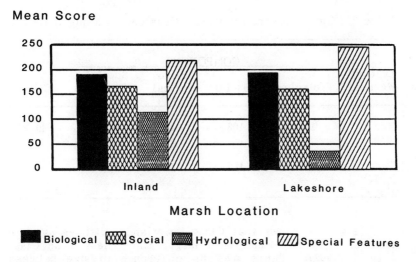

Figure 3. Southern Ontario marsh evaluation scores.

and lakeshore swamps and between inland and lakeshore marshes. However, no significant differences in hydrology were found between different types at the same physiographic site (i.e., between swamps and marshes located inland nor between those located on a lakeshore). A k-sample median test showed no differences in the special features component between these four wetland groups.

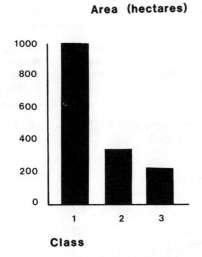

Figure 4. Average wetland area by wetland class (n = 144).

Table 3. Median scores for wetland evaluation components.

| Type of Wetland | COMPONENT | | | | |
	Biological	Social	Hydro-logical	Special Features	SCORE
Marsh (n=69)	196	163	38	250	640
Swamp (n=75)	192	159	119	236	662

Figure 4 shows that Class 1 wetlands, with an average area of 1,000 ha, were much larger than Class 2 or Class 3 ($p < 0.001$). There was no difference in size between Class 2 and Class 3 wetlands.

Because median special features scores were consistently higher than those of the other components (Table 3), these subcomponents were analyzed to examine the differences between the two main wetland types. Differences for four special features values for which high scores were possible are noted in the table below (Table 4).

Table 4. Special features scores for marsh and swamp—subcomponent values with significant differences.

Subcomponent	Marshes Mean Score (n=75)	Swamps Mean Score (n=69)	Probability that scores are not different
Breeding or feeding habitat for a provincially signifi-cant animal species	136.7	97.8	< 0.0001
Winter cover for wildlife	6.5	18.9	< 0.0001
Waterfowl staging	34.7	12.3	0.0005
Significance for fish spawning and rearing	22.3	13.4	0.0014

Additional values under the special features component which were not important in scoring differences between marshes and swamps include: use by migratory passerines and shorebirds, waterfowl production, the presence of provincially significant plants and regionally significant species.

DISCUSSION

Hydrology

Collins and Maltby (1984) observed low hydrological scores for lacustrine marshes along Lake Ontario. The present study shows that the difference in a wetland's hydrological value is clearly influenced more by its site (i.e., whether it is inland or lakeshore) than its type (i.e., whether it is a marsh or a swamp). The value for water detention and flow stabilization is less for wetlands located on the Great Lakes than for wetlands located hydrologically distant. The evaluation procedure places considerable importance on a wetland's role in flow stabilization. (The maximum possible score for this attribute is 190 points.) In the evaluation process wetlands located on the Great Lakes shoreline are not eligible for these points. Inland wetlands scored an average of 81.0 points for flow stabilization.

As most of the swamps were inland (62 of 69), this explains the initial perceived difference in hydrological value between swamps and marshes. The results obtained from the wetlands evaluation procedure clearly indicate no inherent difference between the overall hydrological value of swamps and marshes.

Special Features

Significant loss of wildlife occurs with degradation or elimination of marsh habitat (Jaworski, 1978, 1979; Lemay, 1980). Several important subcomponents of the special features component scored significantly higher for marshes than for swamps (Table 4). Animals considered provincially significant in the wetland evaluation procedure occur more often in marshes than in swamps. Provincially significant birds occurred 187 times in 69 marshes and only 87 times in 75 swamps. (Provincially significant mammals, herptiles and fish occurred much less often than did provincially significant birds.) The few remaining Lake Erie wetlands particularly represent the more important waterfowl staging areas in southern Ontario (McCullough, 1981). Marshes are also important in fish spawning and rearing and several workers have reported that the larval stage when feeding begins is critical to the survival of year classes (Cushing, 1974; Duffy and Liston, 1984). Further studies have been called for in this important area (Herdendorf et al., 1981).

Winter cover for wildlife scored considerably higher for swamps than for marshes. This is undoubtedly a function of the greater number and higher levels of vegetational strata present in swamps. Forest and thicket swamps are the most common wetland type in southern Ontario. The value for wildlife of this wetland habitat has been discussed for the southeastern United States (Wharton et al., 1981), but intensive studies in southern Ontario are still lacking.

Size is also a critical factor in the ranking of wetlands according to the wetland evaluation procedure, irrespective of whether a wetland is marsh or swamp. It may be readily seen that winter cover for wildlife (especially deer), use by migratory waterfowl, and fisheries values are dependent on sufficient size of habitat.

Assessment of the types and characteristics of provincially significant wetlands is important in providing for their long-term management and protection.

ACKNOWLEDGMENTS

Thanks are extended particularly to C. Wedeles for statistical analysis and manuscript preparation. Assistance from P. Wilford and B. Parker for technical assistance was also much appreciated.

LITERATURE CITED

Bardecki, M.J.   1981.   Wetland conservation policies in Southern Ontario: A delphi approach. Unpublished Ph.D. Thesis for Department of Geography, York University, Toronto, Ontario.

Collins, B. and L. Maltby.   1984.   A statistical analysis on "An Evaluation System for Wetlands of Ontario, First Edition (1983)."   Environment Canada, Canadian Wildlife Service, Ontario Region, 32 pp.

Cox, E.T.   1972.   Estimates of cleared wetlands in Southern Ontario.   Unpublished paper prepared for the Wildlife Branch, Ontario Ministry of Natural Resources, Toronto, Ontario.

Cushing, D.H.   1974.   The possible density-dependence of larval mortality and adult mortality in fishes. In: The Early Life History of Fish, J.H.S. Blaxter (ed.),

The Proceedings of an International Symposium held at Dunstaffrage Marine Research Laboratory of the Scottish Marine Biological Association at Oban, Scotland, May 12-23, 1974. Springer-Verlag, New York, pp. 103-111.

Duffy, W.G. and C.R. Liston. 1984. Zooplankton production and zooplankton-larval fish interactions in the Pentwater Marsh on Lake Michigan. Proposal to Michigan Sea Grant, Michigan State University, Dept. Fish. Wild., 8 pp.

EC/OMNR. 1983. An evaluation system for wetlands of Ontario south of the Precambrian Shield, first edition. Canada/Ontario Steering Committee on Wetland Evaluation, Environment Canada and Ontario Ministry of Natural Resources. Part I. Rationale, 50 pp. Part II. Procedures Manual, 74 pp. Part III. Data Record, 15 pp. Part IV. Evaluation Record, 5 pp.

EC/OMNR. 1984. An evaluation system for wetlands of Ontario south of the Precambrian Shield, second edition. Canada/Ontario Steering Committee on Wetland Evaluation, Environment Canada and Ontario Ministry of Natural Resources. Part I. Rationale and Procedures, 72 pp. Part II. Scoring, 21 pp. Part III. Data Record, 25 pp. Part IV. Evaluation Record, 6 pp.

Glooschenko, V. 1983. Development of an evaluation system for wetlands in Southern Ontario, presented at Second Annual Meeting of the Society of Wetland Scientists, St. Paul, MN. Wetlands 3:192-200.

Glooschenko, V., B. Parker, L. Coo, R. Kent, C. Wedeles, and J. Dawson. 1985. Provincially and regionally significant wetlands in Southern Ontario. Interim Report. Ontario Ministry of Natural Resources and World Wildlife Fund, Canada, Queen's Park, Toronto, Ontario, 280 pp.

Green D. 1984. Factors related to assignment of class rank in the Ontario Ministry of Natural Resources Wetland Evaluation System. Unpublished report for Wildlife Branch, Ontario Ministry of Natural Resources, Queen's Park, Toronto, Ontario.

Herdendorf, C.H., S.M. Harley, and M.D. Barnes (Eds.). 1981. Fish and Wildlife Resources of the Great Lakes Coastal Wetlands within the United States. Volume One: Overview. U.S. Fish and Wildlife Service, Washington, DC.

Jaworski, E. and C.N. Raphael. 1978. Coastal wetlands value study in Michigan, Phase I: Fish, wildlife and recreational values of Michigan's coastal wetlands. Twin Cities. U.S. Fish and Wildlife Service.

Jaworski, E. and C.N. Raphael. 1979. Mitigation of fish and wildlife habitat losses in Great Lakes coastal wetlands. In: The Mitigation Symposium, pp. 152-156. Fort Collins, U.S. Department of Agriculture.

Lemay, M.H. 1980. Assessment of the effects of urbanization as a basis for the management of the waterfront marshes between Toronto and Oshawa, Ontario. Waterloo. M.A. Thesis, University of Waterloo.

Lynch-Stewart, P. 1983. Land use changes on wetlands in Southern Canada: Review and bibliography. Working Paper No. 26, Lands Directorate, Environment Canada, Ottawa, Ontario.

McCullough, G.B. 1981. Wetland losses in Lake St. Clair and Lake Ontario. Proceedings of the Ontario Wetland Conference. Anne Champagne (Ed.), 18-19 September 1981. Toronto, Ontario, pp. 81-89.

Reid, R. 1979. Shrinking wetlands: What you can do. Ontario Naturalist 19(2):38-41.

Reid, R. 1981. A critics view of wetland policies. In: Proceedings of the Ontario Wetlands Conference. Anne Champagne (Ed.), 18-19 September 1981. Toronto, Ontario, pp 98-109.

SAS Institute, Inc. 1982. Statistical Analysis System. Cary, NC.

Snell, E.A. 1985. Wetland distribution, losses and mapping approach for Southern Ontario. Lands Directorate, Environment Canada.

Sokal, R.R. and F.J. Rohlf. 1981. Biometry (2nd Ed.), W.H. Freeman and Co., San Francisco, CA.

Wetland mapping series (second approximation). 1983. Lands Directorate, Environment Canada, 130 pp.

Wharton, C.H., B.W. Lambour, S. Newson, P.V. Winger, L.L. Gaddy, and R. Manke. 1981. The fauna of bottomland hardwoods in southeastern United States. Pp. 87-160 in: Wetlands of Bottomland Hardwood Forest. J.R. Clark and J. Bendorado (Eds.), Dev. Argic. and Managed-Forest Ecology 11, Elsevier, Amsterdam, 401 pp.

Whillans, T.H.   1979.   Historic transformations of fish communities in three Great Lakes bays.   J. Great Lakes Res. 5(2):195-215.

CHAPTER 13

WETLAND THREATS AND LOSSES IN LAKE ST. CLAIR

Gary B. McCullough
Canadian Wildlife Service
London, Ontario N6E 1Z7

ABSTRACT

In Ontario, south of James Bay, the most extensive and highest quality habitat for migrating waterfowl is provided by the shoreline marshes of Lakes Erie and St. Clair. Canadian Wildlife Service studies have shown that the wetlands associated with the eastern shore of Lake St. Clair are presently the most important Ontario staging areas for mallards, black ducks, Canada geese and tundra swans. From 1965 to 1984, 30 percent of the privately owned marshland along the eastern shore of Lake St. Clair has been destroyed--a loss of 1,064 ha. Drainage for agriculture accounted for 92 percent of the loss. Canadian Wildlife Service studies have shown a 79 percent decline in the use of this area by true marsh-dwelling waterfowl during the spring and 41 percent decline in the autumn. In 1984, a new and greater threat to the remaining marshland emerged-- property tax reassessments. These Provincially-administered reassessments have resulted in tax increases of 65 percent and higher on marshland. If the same property were drained and farmed, the taxes would be about half as much, and government tax subsidies would be available to further reduce the cost to the landowner.

Pressure to convert these valuable marshes to agricultural land combined with the recent property reassessment and dramatic increase in taxes will only work against the efforts of the Canadian Wildlife Service and others to protect and preserve the wetlands of Lake St. Clair. More marshes will be destroyed and converted to farmland. North American waterfowl will suffer.

INTRODUCTION

Ontario's wetlands--marshes, swamps, fens and bogs--are disappearing! Various researchers conservatively estimate that 50 to 60 percent of southern Ontario's original wetland area has been lost to development. The real figures are likely much higher.

About 35 percent of the wetlands along the Canadian shorelines of Lakes St. Clair, Erie and Ontario have been lost in the past to various types of development. Today, there remain some 33,000 ha of wetland along the shoreline of these three lakes. These wetlands are, and have been, under development pressure from agricultural, residential, industrial and recreational interests.

Each of the lakes has experienced different types and amounts of development pressure. Lake Erie wetlands have not experienced major alterations except in the vicinity of Point Pelee-Hillman Creek, and these were a result of agricultural interests. Lake Ontario has lost approximately 83 percent of the original marshland present in the heavily populated western basin (McCullough, 1981). Only 650 ha remain today largely because of urban industrial development. Lake St. Clair is a different story again. Wetland threats and losses in Lake St. Clair will be examined to provide a specific example of the pressures that Great Lakes coastal wetlands are, and have been, experiencing and to illustrate the effects that such development pressures can have on some of the wildlife values of these areas.

LAKE ST. CLAIR

Located in extreme southwestern Ontario, Lake St. Clair is surrounded on the south and east sides by thousands of acres of highly productive Class 2 agricultural land. Much of the Lake St. Clair shoreline area was formerly wetland and in response to water level fluctuations of Lake St. Clair, much of the large wetland area would have varied between wet meadow and dry meadow. Historically, most of the meadow environments have been drained and cleared for agriculture. It has been estimated that in Dover Township, along the east side of the lake, only approximately 10 percent of the original wetland acreage remains. Rutherford (1979) documented that between 1910 and 1978 there was a 39 percent decline in Lake St. Clair marshland. Of more concern than the historical losses is the more recent, and ongoing, destruction of the remaining Lake St. Clair wetlands.

In Ontario, south of James Bay, the most extensive and highest quality habitat in terms of use by migratory waterfowl is provided by the large shoreline marshes associated with Lakes Erie and St. Clair. Based on Canadian Wildlife Service studies carried out since the 1960s, the Lake St. Clair marshes rank a close second to the Long Point marshes of central Lake Erie as the most important staging area in southern Ontario, in terms of use by waterfowl during migration periods. Peak numbers at Lake St. Clair have been estimated at 60,000 in the spring and nearly 150,000 birds in the autumn. The wetlands associated with the eastern shore of Lake St. Clair are presently the most important migration staging area in southern Ontario for mallards (Anas platyrhyncos), black ducks (A. rubripes), Canada geese (Branta canadensis), and tundra swans (Olor columbianus), (Dennis et al., 1984). Significant portions of the North American populations of canvasbacks (Aythya valisneria), redheads (A. americana) and tundra swans also utilize the Lake St. Clair marshes during migration periods.

In conjunction with waterfowl surveys and an International Joint Commission Study, the Canadian Wildlife Service in 1978 examined recent losses of privately-owned wetlands in Lake St. Clair (McCullough, 1981). That work has been updated to 1984. Conditions were determined from topographic maps, aerial photos and aerial reconnaissance flights.

Results are presented in Table 1, which provides an area breakdown based on whether the wetland is diked or open to Lake St. Clair. A total of 1,064 ha of valuable wetlands were destroyed and converted to alternate uses

Table 1. Wetland loss in eastern Lake St. Clair (not including Walpole Island) 1965 to 1984.

	Wetland Present			
	1965 (ha)	1984 (ha)	Lost (ha)	Lost (%)
Diked	2,064	1,081	983	48
Open (to Lake St. Clair)	1,510	1,429	81	5
TOTAL	3,574	2,510	1,064	30

Table 2.   Wetland loss along the privately owned eastern
shore of Lake St. Clair, 1965 to 1984.

Wetland	Area (ha)	Diked or Open	Reason for Alteration
1. Thames River	59	diked	agriculture
2. mouth of Thames River	115	open	marina/cottage
		diked	agriculture
3. Bradley Marsh	327	diked	agriculture
4. Balmoral Marsh	11	diked	agriculture
5. Snake Island Marsh	156	diked	agriculture
6. St. Luke's Bay	22	diked	agriculture
7. Patrick's Cove	60	diked	agriculture
8. Mitchell Bay	7	open	marina
9. Mud Creek Marsh	307	diked	agriculture

between 1965 and 1984. That represents a loss of 30 percent
of the privately-owned wetlands that were present along
the eastern shoreline of Lake St. Clair in 1965.

Figure 1 delineates as black areas the wetlands
that have been destroyed since 1965. Wetlands that are
still present are cross-hatched. Drainage for agriculture
accounted for 92 percent of the wetland loss, while recreational
marina/cottage development accounted for the other 8 percent
(Table 2). Three agricultural drainages (the Bradley,
Snake Island and the Mud Creek Marshes) account for 74
percent of the total loss. These once productive marshlands
now grow corn.

During the early 1970s, high water levels in Lake
St. Clair temporarily eliminated another 800 ha of emergent
shoreline vegetation in the St. Luke's Bay area, indicated
by the stipled area. This temporary loss of additional
emergent marsh has made the effects of the elimination
of 1,064 ha of marshland more severe for wildlife species
dependent upon wetlands.

In the large Walpole Island marshes contained in
the Walpole Island Indian Reservation to the north of
Chenal Ecarte, smaller recent losses have taken place,
both in terms of actual acreage and percent loss. From
1963 to 1984, 485 ha of wetland were drained for agricultural
purposes. This represents a 5 percent loss from the 11,000

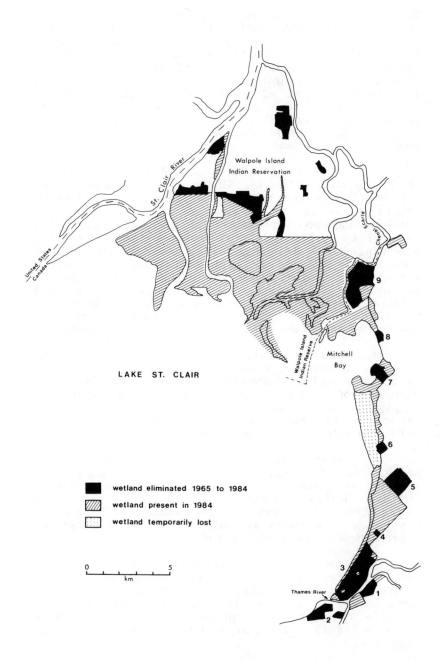

Figure 1. Wetland losses at Lake St. Clair, 1965 to 1984.

ha present in 1963. While the percentage of wetland loss is not great, during that same time period many acres of wetland were enclosed by dikes to enable management by control of water levels. As illustrated by the losses south of Chenal Ecarte, diking of wetlands can be the first step toward drainage for alternate uses. Along that eastern shore, 92 percent of the wetland that was destroyed was diked. This is not to say that the diking of wetlands is necessarily bad. Many productive wildlife management techniques can be employed in diked, water-level controlled wetlands.

Reduced wetland habitat should result in reduced use by migrating waterfowl. Studies by the Canadian Wildlife Service (Dennis and North, 1984) have shown a dramatic decline in the use of the Lake St. Clair marshes by true marsh-dwelling species such as American wigeon (*Anas americana*), gadwall (*A. strepera*), green-winged teal (*A. crecca*), blue-winged teal (*A. discors*) and wood duck (*Aix sponsa*). From 1968 to 1982 the use by these dabbling ducks declined a dramatic 79 percent during the spring migration. Autumn use declined 41 percent. Spring use generally provides a better indication of the value of the wetlands to migrant waterfowl, since management techniques, used during the fall hunting season are not implemented during the spring.

It should be pointed out that even though use by true marsh-dwelling species declined dramatically, use by *total* waterfowl increased in the Lake St. Clair area. A number of factors are responsible, but the primary one is population increases in mallards and Canada geese, two species that respond well to man's management techniques and do not rely heavily on large areas of wetland habitat.

At present, all wetlands existing in Lake St. Clair are maintained primarily as private duck hunting clubs, with the exception of the 240 ha Canadian Wildlife Service St. Clair National Wildlife Area. On the Walpole Island Indian Reservation, most of the marshland not leased to hunt clubs is operated as public hunting areas under the direct control of the Walpole Island Band.

It is an expensive proposition to own and manage these private marshlands, and the expense will continue to increase. The pressure to convert these marshes to farmland, or to sell the land to farming interests, will continue at least at the present level. That pressure has resulted in a 1.6 percent loss of wetlands per year since 1965.

THE LATEST THREAT

In 1984, a new and greater threat to the remaining Lake St. Clair marshland emerged on the scene, property tax reassessments.

Under the newly implemented Section 63 Reassessment Program, requested by the local Municipality and administered by the Provincial Ministry of Revenue, all land in Dover Township on the east side of Lake St. Clair was reassessed using 1980 market values. Each property was assigned to one of four property classes: farmland (2.1%), residential (4.5%), commercial (7.5%), industrial (8.8%). The property class percentage figure is applied to the market value to produce an assessed value. The taxes payable are arrived at by multiplying the assessed value by the mill rate.

The "kicker" is that marshalnd is considered to be recreational land, which is included in the "residential property class" and factored at 4.5 percent, more than twice that for farmland. No appropriate property class for marshland exists.

As a result of the reassessment, 1984 taxes on the Canadian Wildlife Service's 240 ha St. Clair National Wildlife Area have jumped by $5,000, a 65 percent increase. The taxes on nearby, privately-owned marshes have increased by at least as much if not a higher percent. While the Canadian Wildlife Service objected to and opposed the reassessment of the National Wildlife Area, of much more concern is the general reassessment of all marshland associated with the eastern shore of Lake St. Clair.

An interesting sidepoint is that the Township of Dover has passed a zoning bylaw which recognizes the importance of the remaining Lake St. Clair marshes. This bylaw designates the Lake St. Clair shoreline wetlands as MARSHLAND, and the zoning restricts the permitted uses of these lands. The Township is to be commended. However, the reassessment of these lands as "residential" does not recognize the marshlands, and such a designation is at odds with the Township zoning. On one hand, a landowner is restricted by the bylaw in terms of permitted land uses, but on the other he is taxed as if he could or did use the property for true residential land.

From a property tax standpoint alone, the pressure to drain marshes and convert to farmland is now greater. This is especially so since generous tax subsidies for farmland are available from the Provincial government. No such subsidies exist for wetlands.

As previously stated, 30 percent of the privately-owned marshland along Lake St. Clair's eastern shore has been destroyed since 1965. Drainage for agriculture accounted for 92 percent of that loss of 1,064 ha of valuable marshland.

Despite the economic value of waterfowl hunting in the area, the future of many important marshes along the eastern shore of Lake St. Clair is uncertain. The recent reassessment and dramatic increase in taxes will only work against the efforts of the Canadian Wildlife Service and others to preserve and protect the wetlands of Lake St. Clair. The increased taxation will likely result in the conversion of even more Lake St. Clair marshes to agricultural land. North American waterfowl populations will suffer.

I agree with many of the previous speakers that much more research is required to understand these complex wetland ecosystems. However, unless greater effort from all fronts is directed at finding more effective means to preserve and protect the remaining Great Lakes coastal marshes, there will be little or nothing left to understand.

LITERATURE CITED

Dennis, D.G. and N.R. North. 1984. Waterfowl use of the Lake St. Clair marshes during migration in 1968-69, 1976-77, and 1982. Pages 43-52 in: S.G. Curtis, D.G. Dennis and H. Boyd (eds.), Waterfowl Studies in Ontario, 1973-81. Can. Wildl. Serv. Occ. Paper No. 54, 71 pp.

Dennis, D.G., G.B. McCullough, N.R. North, and R.K. Ross. 1984. An updated assessment of migrant waterfowl use of the Ontario shorelines of the southern Great Lakes. Pages 37-42 in: S.G. Curtis, D.G. Dennis and H. Boyd (eds.), Waterfowl Studies in Ontario, 1973-81. Can. Wildl. Serv. Occ. Paper No. 54, 71 pp.

McCullough, G.B. 1981. Wetland losses in Lake St. Clair and Lake Ontario. Pages 81-89 in: A. Champagne (ed.), Proc. of the Ontario Wetlands Conference, Toronto, Ontario, 18-19 Sept. 1981, 193 pp.

Rutherford, L.A. 1979. The decline of wetlands in southern Ontario. Faculty of Environmental Studies, University of Waterloo, Waterloo, Ontario.

CHAPTER 14

HUMAN INTERFERENCE WITH NATURAL WATER LEVEL
REGIMES IN THE CONTEXT OF OTHER CULTURAL
STRESSES ON GREAT LAKES WETLANDS

Nancy J. Patterson
Federation of Ontario Naturalists
Don Mills, Canada   M3B 2W8
and
Thomas H. Whillans
Trent University
Peterborough, Canada   K9J 7B8

INTRODUCTION

Water level regime is but one of many manageable factors which could influence the condition or extent of a Great Lakes wetland. Some factors which could affect water levels such as river discharge into a wetland, diversion of lake water around a wetland, isolation from natural hydrologic influence (dyking) or channelization through a wetland could also have independent influence and are subjects of considerable human tampering. It is therefore advisable to consider water level regime and human interference with it in the context of other human-engendered problems in Great Lakes wetlands. There are at least three major aspects which merit examination:

(1)   Comparison of causal factors in order to isolate similarities among causes (and implied solutions);

(2)   Contrast of stresses (biological, chemical or physical perturbation) and of long-term responses in order to clarify the ecosystemic significance of water level regime (and implied priority for action); and

(3)   Investigation of interaction among causes, among stresses and among long-term responses in order

to specify synergisms and antagonisms (and implied interpretation of (1) and (2)).

The aspects (1) and (2) have been examined to a degree for the Great Lakes in general (Francis et al., 1979; Whillans, 1980), for certain wetland-rich ecosystems within the Great Lakes (Harris et al., 1982); Francis et al., 1985), and for wetlands in general (Patterson, 1984). This review is based in large part upon those studies.

It is unlikely that (3) can be addressed in a comprehensive manner at the moment. It could, however, be very important. For example, the remarkable natural water level fluctuation in the Great Lakes coupled with the high wave energy characteristic of much of the Great Lakes shoreline suggests that distribution of wetlands in the Great Lakes will be limited to protected pockets and that long-established wetland biota will be adapted to substantial water level fluctuation. In fact, these two factors are related. Thus, biota which colonize isolated morphometrically protected wetlands must have some way of surviving water level extremes which temporarily render the wetlands inhospitable. For example, Keddy and Reznicek (see Chapter 3, this volume) have demonstrated how seed banks represent a strategy of suspending plant productivity during inhospitable conditions, given that hospitable conditions will eventually re-occur. Johnson and Snyder (1984) have shown how the opportunity for fish to move readily into and out of a wetland appears to influence the capacity of that wetland to support a wider variety of fish. The implication is that fish mobility and opportunity for refuge would facilitate response to water level extremes.

There are other adaptations, and the degree to which species rely on any of these will vary among species. The important point to make here, however, is that normal wetland response to water level fluctuation is likely tied to the timing and degree of those fluctuations, other characteristics of the wetland, and conditions external to the wetland. Any stress, be it natural or cultural, which influences any of those factors could modify wetland response to water levels.

WETLANDS, WATER LEVELS AND STRESSES IN GENERAL

All wetlands in the Great Lakes are subject to stresses of some sort. Along the United States shoreline, there are almost 121,500 hectares of wetland (Herdendorf and Hartley, 1979). Comparable figures are unavailable for the Canadian shoreline. However, wetlands represent almost

Table 1. A taxomony of cultural stresses occurring in the Great Lakes.*

Natural background processes	Battering storms; rains and floods; water cycles; spells of hot or cold weather; forest and marsh fires; disease outbreaks.
Harvesting of renewable resources	Fishing whether commercial, sport or domestic; hunting for ungulates, upland birds or waterfowl; trapping for muskrat or fox; withdrawal of water for consumption.
Loading by substances and heat energy	Inert solids and suspensions of sand and clay; nutrient materials that fertilize plants and plankton; poisons that kill organisms; contaminants that affect health of organisms; heat that raises the temperature of the water.
Restructuring the morpho-metric form of water bodies	Filling in deeper parts with sediments; damming streams; modifying the shoreline by bulkheading, infilling, etc.; dredging to deepen parts of the basin; stirring up bottom by boating and shipping.
Introduction of non-native organisms	Intentional stocking of preferred organisms which may nevertheless become pests; accidental invasion via canals; accidental introduction via bilge water, private aquaria, anglers' bait buckets, etc.

*Extracted from Francis et al. (1985).

1,500 of roughly 4,500 km of mainland shoreline length in Canada (Whillans, 1984).

A taxonomy of cultural stresses occurring in the Great Lakes is provided in Table 1. This represents major classes of stress, each of which represents a number of specific stresses. The most pronounced stresses on wetlands have historically occurred in the lower Great Lakes and southern Lake Michigan where human activity has been most intensive. These are also the regions of largest wetland area. The lower Great Lakes contain some 60,750 hectares of wetland (Lake Erie Water Level Study, 1981); Lake Michigan has about 48,600 hectares (Herdendorf and Hartley, 1979). This represents current wetland area.

Table 2. Summary of wetland losses in the lower Great Lakes and Lake Michigan.

Location	Historic Period	Magnitude of loss (% of former area)	Authority
Lake Erie and Detroit River, United States	1916 to 1967-73	61	Jaworski and Raphael, 1978
Lake St. Clair, United States	1873 to 1973	72	Jaworski and Raphael, 1978
Saginaw Bay, Lake Huron	1956 to 1963-73	53	Jaworski and Raphael, 1978
Little Bay de Noc, Lake Michigan	1910 to 1958	50	Jaworski and Raphael, 1978
Les Chemeaux, Lake Michigan	1904 to 1964	57	Jaworski and Raphael, 1978
Lake St. Clair, Canada	1965 to 1978	25	McCullough, 1981
Niagara River to Toronto, Lake Ontario	1800s to 1976	73-100	Whillans, 1982
Toronto to Bay of Quinte, Lake Ontario	1800s to 1976	8-32	Whillans, 1982

Losses since historic times would indicate something about gross long-term impact of stress. McCullough (see Chapter 13, this volume) has reported on losses in Lake St. Clair. Table 2 summarizes some estimates of losses. Jaworski and Raphael (1978) have shown that natural water level fluctuations caused a significant proportion of the losses either through direct impact on vegetation or by accelerating erosion of protective barrier bars. The losses in Les Chemeaux Islands, for example (Table 2), were mostly induced by natural water level fluctuation. In other locations, however, areal gains were attributed to lake levels. The other authors in Table 2 acknowledged the role of natural lake levels; however, like Jaworski and Raphael (1978), they indicated that in areas of intensive urban, recreational or agricultural activity the impacts of other stresses were overwhelming. Thus, losses in southern Lake Michigan, Saginaw Bay, Lake St. Clair, western Lake Erie, and western Lake Ontario constitute losses to gross stress, mostly cultural. Moreover, changes in species composition, density, etc., were often more remarkable than areal loss.

Of course, many Great Lakes wetlands are somewhat independent of the natural Great Lakes' water level regime and problems of lacustrine origin. Some 18,225 of the 60,750 hectares of wetland in the lower Great Lakes are morphometrically isolated from the water level regime of the Great Lakes. Almost two-thirds of that acreage is along the United States' shoreline, representing the large extent of dyking that has taken place there. Over half of the United States wetland area is now morphometrically isolated.

In summary, then, changes in wetland area are very crude indicators of stress in that:

(1)   wetland area can be expected to increase and decrease through time in response to natural water levels; the direction of change measured would partly be a function of the interval of measurement;

(2)   highly stressed wetlands may respond qualitatively, rather than in areal extent; and

(3)   expansion of wetland area is in many cases morphometrically precluded.

Natural climatic influences on water levels predominate in all of the Great Lakes; in Lakes Ontario and Superior some human dampening of extremes has been engineered; in many wetlands a significant degree of artificial control of levels has been achieved; and for all of the Great Lakes the potential for large scale alteration of water level regime must be recognized. Given the variety of ways in which water level regime may be influenced, the assortment of wetland responses, and the number of other stresses; it is both difficult and necessary to begin to develop a framework for relating those factors.

STRESSES, THEIR CAUSES AND RECOGNITION

The two most numerous broad types of wetland in Great Lakes coastal waters are marshes and swamps. Each

of these may be subdivided into ecologically more definitive
categories, for example: deltaic, morphometrically isolated,
and lakeside. The categories do, however, have many similarities
and for the purposes of this discussion may be grouped
together.

All of the classes of stress presented in Table
1 could in some way involve marshes and swamps. Some,
however, are more frequently associated. Table 3 presents
15 cultural stresses that are predominant in Great Lakes
wetlands. For each stress, the predominant causal agents
are indicated.

Many of the stresses in Table 3 may have similar
causes. It is evident that some activities such as channelization
and spoil deposition have potential to cause many types
of stress. It should also be apparent that different
stresses could result in similar changes in wetland characteristics. This is detailed in Table 4 for marshes and swamps.

The changes in wetland characteristics--the symptoms
associated with each stress--have been grouped into (1)
physical/chemical changes, (2) vegetation changes and
(3) ecological changes; but these are not mutually exclusive.
The sources of information on the stresses and symptoms
of stresses were not limited to the Great Lakes literature.
They should, however, apply in principle if not in degree.
In some cases, inferences were made from literature on
wetland types other than marshes or swamps.

A complicating factor is that some of the symptoms
of stresses listed in Table 4 can themselves be stresses.
In Table 5 the initial 15 cultural stresses are presented
in conjunction with 15 secondary stresses which they may
induce or at least predispose a wetland towards. In other
words a sequence of stresses and responses could ultimately
result from a rather simple cultural activity.

While it is valuable to be able to recognize and
identify symptoms of disturbed wetlands, to predict cause
from symptoms is not possible for a variety of reasons:

(1)   A cause and effect relationship implies a
level of understanding of wetland ecology that may not
have been attained. Investigations are still at a rudimentary
stage.

(2)   Each situation is unique. The ability of
a system to withstand or respond to stress depends on,
among other things, its vulnerability, inertia, elasticity
and resiliency (Cairns and Dickson, 1977). Diverse combinations

of stresses and abiotic components of the ecosystem will further influence the response pattern. Generalizations based on single stresses and generalized wetland types can therefore only provide examples of possible responses to a stress. The symptoms listed will not necessarily be observed in every stress-response interaction.

(3)   The question of intensity and spatial extent of stress tends to be neglected. For example, symptoms listed for a marsh in which the water depth has been increased are the same regardless of the flooding depth. Yet a marsh flooded with 2 meters of water may not have the same symptoms as one flooded with 0.5 meters.

Another example involves nutrient loading. With increased influx of nutrients, particularly nitrogen and phosphorus, primary productivity of a wetland is stimulated. As nutrients continue to be pumped into the system, the wetland reaches a threshold beyond which respiration exceeds the oxygen transport, and the system becomes anoxic. Although the type of stress is unchanged, the wetland can evolve from a stage of enhanced production to one in which aquatic conditions are markedly degraded (Darnell et al., 1977).

(4)   Many stresses result in the same or similar responses. When interpreting symptoms, it is often difficult to isolate one causal agent. Increased turbidity, for example, may suggest sediment loading, channelization of flow, increased water level or any one of a number of combinations of other stresses.

(5)   A combination of stresses can result in an unexpected response if only the predicted symptoms for the given stresses are considered. The combined effect of stresses may produce a "synergistic" ecosystem perturbation—one that is greater than the sum of the individual stresses. On the other hand, a combination of stresses may result in one stress ameliorating an effect of another stress. The result would be an "antagonistic" perturbation--one that is somewhat less than the sum of the individual stresses. Synergism and antagonism are particularly important concepts when dealing with toxic loading. As mentioned at the beginning of this paper, stress interactions are not well understood.

CONTROL OF STRESSES

It is not possible at the moment to make a definitive or comprehensive conclusion on the stress regime with respect to Great Lakes wetlands. This paper has presented

Table 3. Primary causal agents of selective cultural stresses.[1]

Selective Cultural Stresses	Primary Causal Agents														
	DC	DA	DI	DR	FI	DD	CH	SD	GS	VD	WW	RO	GC	HA	NA
HYDROLOGIC FLOW MODIFICATION															
Increase in water depth	*	*					*		*				*		
Decrease in water depth	*	*				*	*						*		
Abnormal fluctuations in depth			*	*	*	*									
Reduction in lateral flow	*		*	*	*		*	*					*		*
Channelization of lateral flow				*									*		
WATER QUALITY DEGRADATION															
Nutrient loading (N and P)	*	*		*		*					*	*	*		*
Toxic loading	*	*		*		*					*	*	*		*
Thermal loading										*					
Sedimentation	*	*		*	*		*	*		*			*		*
Turbidity	*	*		*	*		*	*		*			*		*
ECOLOGICAL STRUCTURAL BREAKDOWN															
Bottom configuration				*		*							*		*
Biomass export														*	

Table 3. Continued.

| | Primary Causal Agents | | | | | | | | | | | | | | |
Selective Cultural Stresses	DC	DA	DI	DR	FI	DD	CH	SD	GS	VD	WW	RO	GC	HA	NA
ECOLOGICAL STRUCTURAL BREAKDOWN (Cont.)															
Pests								*							
Fire						*									
Ice Damage															

[1]Source: Patterson (1984)
[2]Key:
DC = Dyke construction
DA = Damming with utility corridors
DI = Ditching
DR = Dredging
FI = Filling
DD = Drawdown
CH = Channelization
SD = Spoil deposition
GS = Generating station
VD = Vegetation destruction
WW = Discharge of wastewater or municipal sewage
RO = Runoff or groundwater
GC = General construction
HA = Harvesting
NA = Navigation

Table 4. Symptoms observed in two types of freshwater wetlands affected by selective cultural stresses.[1]

SYMPTOMS	REFERENCES
STRESS: INCREASE IN WATER DEPTH HYDROLOGIC FLOW MODIFICATION	
Marsh	
A. Physical/Chemical changes	
1. decreased alkalinity	Lathwell et al., 1969; Kadlec, 1962
2. increased turbidity with loss of wetland fringe	Darnell et al.,1977; Gosselink and Turner,1978
3. increase shoreline and dyke erosion	Harris and Marshall, 1963; Allen, 1979
4. change in wetland size—either increase or decrease	Herdendorf and Hartley, 1979
5. improved circulation important in lakeshore	Jaworski et al., 1979
6. increased siltation	Jaworski et al., 1979
B. Vegetation changes	
1. decrease in density	Herdendorf and Hartley, 1979; Harris and Marshall, 1963
2. community changes	
-emergent die-off at inland fringe	Harris and Marshall, 1963; Geis, 1979; Millar, 1973
-emergent replaced by submergent in deep water	International Joint Commission, 1980
3. decreased productivity	Geis, 1979; Cook and Powers, 1958; Van der Valk and Bliss, 1971; Van der Valk and Davis, 1978a

C. Ecological changes
  1. loss of valuable edge habitat, therefore
     decreased nesting, egg hatching, etc.   Harris and Marshall, 1963
  2. food shortage with decrease in insects
     and vegetation   Weller, 1978
  3. change in fish habitat and spawning   Jaworski et al., 1979
  4. decreased muskrat habitat, drowned young   Bellrose, 1950; Weller, 1978
  5. shift in bird community   Jaworski et al., 1979

Swamp
A. Physical/Chemical changes
  1. decrease of soil oxygen and build-up of
     compounds toxic to roots   Harms et al., 1980
  2. decreased alkalinity   Lathwell et al., 1969
  3. increased shoreline and bank erosion   Allen, 1979

B. Vegetation changes
  1. change from Carex spp. to emergents   Andrews, 1977
  2. tree death   Clarke and Clarke, 1979; Harms et al., 1980
  3. herbaceous community—little change or
     may be enhanced   McLeese and Whiteside, 1977

C. Ecological changes
  1. decrease in seed germination   Clarke and Clarke, 1979
  2. decreased aesthetics   Jeglum, 1975
  3. loss of habitat for birds, fish and mammals   Darnell et al., 1977
  4. decreased winter food for some animals   Darnell et al., 1977

Table 4. Continued.

SYMPTOMS	REFERENCES

HYDROLOGIC FLOW MODIFICATION

STRESS: DECREASE IN WATER DEPTH

Marsh

A. Physical/Chemical changes

1. increase in rate of decomposition of organic	Cook and Powers, 1958
2. increase in nutrient release from sediments	Cook and Powers, 1958; International Joint
	Commission, 1980
	Lathwell et al., 1969; Kadlec, 1962
3. increased alkalinity	Kadlec, 1962
4. increase in water temperature	

B. Vegetation changes

1. decreased primary productivity	Clarke and Clarke, 1979
2. increased species diversity	Auclair et al., 1973
3. decrease in number of dominant species	Auclair et al., 1973
4. change in density—some species resistant,	
others stimulated	Cooke, 1980
5. community changes	
-woody shrubs may invade shallow water	International Joint Commission, 1980
-exposure of seed bank and germination of:	Van der Valk and Davis, 1978b; Peterman, 1980
-emergents in very shallow water	
-submerged aquatics in standing water	
-mud flat species in saturated soil	

C. Ecological changes
1. increased frequency of fire and subsequent invasion of non-wetland species or change   Clarke and Clarke, 1979
2. loss or change of habitat for existing fish community   Clarke and Clarke, 1979
3. severe decrease or loss of muskrat population due to frozen routes to feed beds and frozen food   Friend et al., 1964
4. disruption of habitat corridors   Weller, 1978

Swamps
A. Physical/Chemical changes
1. increased nutrient release from sediments   Cook and Powers, 1958

B. Vegetation changes
1. increased tree growth   Found et al., 1974
2. reduction in common wetland plants   McLeese and Whiteside, 1977
3. succession from semi-aquatic to annual weeds   Meeks, 1969
4. reduction in some woody species such as cottonwood and silver maple   Found et al., 1974

STRESS: ABNORMAL WATER DEPTH FLUCTUATIONS (DRAWDOWN)

HYDROLOGIC FLOW MODIFICATION

Marsh
A. Physical/Chemical changes
1. reduced turbidity   Kadlec, 1962; Harris and Marshall, 1963; Cooke, 1980
2. soil and water nutrients increased   Kadlec, 1962;
3. substrate consolidates very hard   Kadlec, 1962; Cooke, 1980
4. aerobic conditions during drawdown   Kadlec, 1962

Table 4. Continued.

SYMPTOMS	REFERENCES
B. Vegetation changes	
1. reduction in dry matter rather than vegetative death	Geis, 1979
2. periphery vegetation denser than inside vegetation	Harris and Marshall, 1963
3. stands developing on coarse alkaline soils less dense than on organic soils	Harris and Marshall, 1963
4. algal blooms after reflooding	Cooke, 1980
5. community changes	
—many submergent and floating leaves may be lost	Kadlec, 1962
—existing communities of emergents spread and increase in abundance	Kadlec, 1962
—summer dry period results in some invasion of terrestrial plants	Cooke, 1980
—invasion of resistant species	Cooke, 1980
—with reflooding, moist species, i.e., cattail, bullrush sedges, etc., colonize regardless of the type of vegetation present at the end of drawdown	Harris and Marshall, 1963
C. Ecological changes	
1. zonation and distribution of vegetation altered after reflooding depending on depth and permanence of restored water	Harris and Marshall, 1962

2. loss of most mollusks and invertebrates   Kadlec, 1962
3. some loss of wildlife
    -fish kills possible   Cooke, 1980
    -muskrat kills if drawdown over winter   Friend et al., 1964

HYDROLOGIC FLOW MODIFICATION

STRESS: REDUCTION IN LATERAL FLOW

A. Physical/Chemical changes
    1. reduced turbidity   Smith, 1970
    2. reduced nutrient and silt input   Clarke and Clarke, 1979
    3. depletion of O_2 and formation of anerobic
        condition   Gosselink and Turner, 1978
    4. accumulation of sediment   Gosselink and Turner, 1978; Clarke and
                                                  Clarke, 1979
    5. increase in soil organic concentration   Grosselink and Turner, 1978
    6. increase in temperature   Darnell et al., 1977

B. Vegetation changes
    1. reduced species diversity   Clarke and Clarke, 1979
    2. reduced primary productivity   Clarke and Clarke, 1979; Gosselink and
                                                  Turner, 1978
    3. reduced plant vigor in some species   Millar, 1973
    4. community changes
        -Carex antherodes or other disturbance-related
         species may invade
        -increase in submergents and semi-aquatics Smith, 1970
         in deeper water
        -may select for a few species at the
         expense of others   Clarke and Clark, 1979

Table 4. Continued.

SYMPTOMS	REFERENCES
C. Ecological changes	
1. formation of a dense senescent wetland	Clarke and Clarke, 1979
2. reduction in quality of habitat for existing community	Clarke and Clarke, 1979

STRESS: CHANNELIZATION OF LATERAL FLOW

HYDROLOGIC FLOW MODIFICATIONS

SYMPTOMS	REFERENCES
A. Physical/Chemical changes	
1. increased erosion (organic erodes faster than mineral)	Stone et al., 1978
2. increased turbidity	Stone et al., 1978
3. reduction in wetland hectareage with spoil disposal	Stone et al., 1978
4. reduction in nutrient and silt deposition if water is shunted through wetland	Patrick, 1978
5. reduced buffering capacity of chemical discharge	Stone et al., 1978
6. rapid widening of channel	Brady, 1979
7. may reduce local groundwater tables	Brady, 1979
8. may increase peak discharge	Larson, 1981
B. Vegetative changes	
1. reduced productivity	Barstow, 1971

2. removal of adjacent vegetation — Clarke and Clarke, 1979, Found et al., 1974

3. vegetation changes similar to sedimentation problem — Brady, 1979

C. Ecological changes
1. reduces quantity and quality of "edge" habitat — Barstow, 1971
2. may reduce environment quality of surrounding water bodies if water purification role of wetlands eliminated — Patrick, 1978; Stone et al., 1978; Found, et al., 1974
3. reduction in number of fish — Groen et al., 1978
4. general reduction in quality of wildlife habitat — Patrick, 1978
5. loss of fish spawning habitat and nursery area — Francis et al., 1979

WATER QUALITY DEGRADATION
STRESS: NUTRIENTS - PRIMARILY NITROGEN (N) AND PHOSPHORUS (P)

Marsh/Swamp
A. Physical/Chemical changes
1. increased organic content — Lee et al., 1975
2. reduction in dissolved oxygen — Lee et al., 1975
3. reduction of nitrate from input water by denitrification — Lee et al., 1975; Kadlec and Kadlec, 1979
4. possible production of tastes and odors — Lee et al., 1975
5. high colour in water — Lee et al., 1975
6. elevated concentrations of N and P in sediments — Kadlec and Kadlec, 1979
7. may act as P source in fall, therefore concentration of P in water is greater leaving wetland than entering — Klopatek, 1978; Lee et al., 1975; Sloey et al., 1978

Table 4. Continued.

SYMPTOMS	REFERENCES
B. Vegetation changes	
1. increased primary production of macrophytes, epiphytes and trees (in swamps)	Dolan et al., 1978; Nicholls and MacCrimmon, 1974; Boyt et al., 1977
2. general reduction in height of vegetation	Whigham and Simpson, 1976
3. nutrient uptake by vegetation	
a. concentrations of many nutrients lower in aged vegetation	Boyd, 1978
b. nutrient concentration in plant tissue decreased seasonally	Prentki et al., 1978
c. increase in concentration of nutrients in both above and below ground vegetation parts	Richardson et al., 1978
d. nutrient enrichment plus high water levels enhances phosphorus storage in some species	Davis and Harris, 1978
e. wetlands with abundant epiphytes may remove more N	Allen, 1971
4. change in vegetation community	
a. some sensitive species may be eliminated	Sloey et al., 1978
b. reduction in diversity of dominant species	Sloey et al., 1978
c. usually increased standing crop of Typha spp.	Hartland-Rowe and Wright, 1975

C. Ecological changes
  1. reduction in suitable fish habitat   Lee et al., 1975
  2. reduction of zooplankton, necton and benthos   Hartland-Rowe and Wright, 1975

STRESS: TOXIC LOADING (heavy metals, synthetics, etc.)

WATER QUALITY DEGRADATION

Marsh/Swamp
A. Physical/Chemical changes
  1. increased concentrations of toxins in sediment   Kadlec and Kadlec, 1979; Aulio, 1980
  2. increased concentrations of toxins in peat   Glooschenko and Capobianco, 1978
    (bogs)
  3. total uptake may be higher with low pH (Cd)   Reddy and Patrick, 1977

B. Vegetation changes
  1. chlorosis on leaves with some metals   McNaughton et al., 1974
  2. some species eliminated, others not visibly   Aulio, 1980; Eriksson and Mortimer, 1975
    affected
  3. elevated concentrations of toxins in plant   Kadlec and Kadlec, 1979; Clarke and Clarke,
    tissue   1979
  4. productivity lower in some species   Clarke and Clarke, 1979
  5. elevated concentrations of toxins in mosses   Glooschenko and Capobianco, 1978;
    of bogs   Gorham and Tilton, 1978
  6. tolerant races may develop, i.e., Nuphar and Cu   Aulio, 1980

C. Ecological changes
  1. recycling of toxins through sediment and   Welsh and Denny, 1976; Cushing Jr., et al.,
    food chains   1980

Table 4. Continued.

SYMPTOMS	REFERENCES
2. bioaccumulation of some toxins through food chains	Welsh and Denny, 1976; Francis et al., 1979
3. high concentrations of toxins may have deleterious effects on animal communities and may cause ecological abnormalities	Welsh and Denny, 1976; Kadlec and Kadlec, 1979; Boto and Patrick, 1979; Clarke and Clarke, 1979

WATER QUALITY DEGRADATION

STRESS: THERMAL LOADING

Marsh/Swamp

A. Physical/Chemical changes

1. increase in overwinter temperature of groundwater	Bedford, 1977, Adriano et al., 1980
2. can increase biological oxygen demand (BOD)	Krenkel and Parker, 1979
3. increase in turbidity	Krenkel and Parker, 1979
4. increased denitrification rate	Richardson et al., 1978; Sloey et al., 1978

B. Vegetation changes

1. general increase in growth rate in some species	Bedford, 1977, Adriano et al., 1978
2. early spring emergence of some species	Bedford, 1977
3. extensive plant mortality in long term	Bedford, 1977
4. variation in flowering and reproductive phenologies	Christy and Sharitz, 1980

5. vegetation community changes
   -removal of certain thermal sensitive species   Adriano et al., 1980
   -increase in number of annuals   Bedford, 1977; Sharitz et al., 1979
   -reduction in number of trees   Sharitz et al., 1979
   -patchy distribution   Christy and Sharitz, 1980
   -dense algal blooms   Krenkel and Parker, 1969; Bott, 1975;
                                                          Patrick, 1971

C. Ecological changes
   1. ice-free areas in winter increase wildlife
      habitat in winter   Hamely and MacLean, 1979
   2. delayed fall dieback disrupts nutrient and
      organic cycling   Bedford, 1977
   3. dense algal blooms reduce water quality
      and aesthetics   Krenkel, 1969; Bott, 1975; Patrick, 1971
   4. change in faunal community depending on
      thermal requirements   Adriano et al., 1980
   5. change in fish community   Cooley, 1971
      -increased risk in deformities   Darnell et al., 1977; Francis et al., 1979
      -invasion of warm water species (carp)   Krenkel and Parker, 1969
      -reduced quality of fish habitat   Krenkel and Parker, 1969
      -reduced "health" of fishes

WATER QUALITY DEGRADATION

STRESS: INCREASED SEDIMENT LOAD--SEDIMENTATION

Marsh/Swamp
A. Physical/Chemical changes
   1. blanketing of substrate   Boto and Patrick, 1979

Table 4. Continued.

SYMPTOMS	REFERENCES
2. increase in concentration of nutrients, heavy metals, pesticides and other toxins in the substrate	Boto and Patrick, 1979; Richardson et al., 1978; GLB Framework Study, 1975
3. increase in elevation of wetland	Gosselink and Turner, 1978; Reed et al., 1977
4. increased probability of flooding	GLB Framework Study, 1975
5. phosphorus sink	Klopatek, 1978
B. Vegetation changes	
1. increase in vegetation density	Boto and Patrick, 1979; Patrick, 1978
2. with moderate degrees of sedimentation, primary production is stimulated	Boto and Patrick, 1979; Richardson et al., 1978; Birch et al., 1980
3. change in macrophyte distribution	Francis et al., 1980
4. large vegetation such as Phragmites most tolerant	Boto and Patrick, 1979
C. Ecological changes	
1. reduction in bottom habitat diversity	
-elimination of number of benthos species or reduction in numbers	Darnell et al., 1977
-changes in community of attached algae	Darnell et al., 1977
-loss of spawning habitat	Found et al., 1974
2. increased probability of toxin stress (see table for toxins)	
3. reduction in fauna diversity	Darnell et al., 1977; Hynes, 1974

WATER QUALITY DEGRADATION

STRESS: INCREASED SEDIMENT LOAD—TURBIDITY

Marsh/Swamp
A. Physical/Chemical changes
1. increased BOD   Boto and Patrick, 1979
2. may be some flocculation of finer sediments
   by plant exudates   Boto and Patrick, 1979
3. reduced light penetration   Kadlec and Kadlec, 1979
4. increased water temperature   Kadlec and Kadlec, 1979

B. Vegetation changes
1. primary productivity reduced   Boto and Patrick,1979; GLB Framework Study,1975
2. change in vegetation community
   —development of rooted macrophytes inhibited   Darnell et al., 1977; Hynes, 1974
   —submergent vegetation reduced due to
    limited light penetration   Darnell et al., 1977

C. Ecological changes
1. alterations to the fish community   Francis et al., 1979; Hynes, 1974
   —reduced effectiveness in predation   GLB Framework Study, 1975; Darnell et al., 1977
   —reduced egg survival   Darnell et al., 1977
   —reduced "health" of fish
2. general reduction in aquatic fauna
   numbers and diversity   GLB Framework Study, 1975; Darnell et al., 1977

Table 4. Continued.

SYMPTOMS	REFERENCES
ECOLOGICAL STRUCTURAL BREAKDOWN	
STRESS: BOTTOM CONFIGURATION	
Marsh/Swamp	
A. Physical/Chemical changes	
1. large amount of erosion and slumping	Patrick, 1978
2. increased turbidity	Patrick, 1978
3. reduction in diversity of bottom habitats	Patrick, 1978
4. can disrupt lateral flow	Darnell et al., 1977
5. may change water chemistry (heavy metals or nutrients from leachate)	Darnell et al., 1977
B. Vegetation changes	
1. invasion of weedy species on unseeded spoil banks	Brady, 1979
2. invasion of terrestrial species	Van der Valk and Davis, 1978b
C. Ecological changes	
1. loss of wetland habitat	Stone et al., 1978
2. provision of den sites for animals, i.e., muskrat, skunk	Anderson, 1948
3. may increase muskrat population	Anderson, 1948
4. reduction in diversity of benthos	Patrick, 1978

STRESS: BIOAMSS EXPORT   ECOLOGICAL STRUCTURAL BREAKDOWN

Marsh/Swamp

A. Aquatic macrophytes

1. removal of phosphorus sink — Welsh et al., 1979; Loucks, 1981

2. interferes with nutrient cycling of nitrogen and phosphorus — Welsh et al., 1979; Loucks, 1981

3. vegetation community changes depending on harvest frequency
   - occasional harvest community shifts to monotypic — Cottam and Nichols, 1970; Loucks, 1981
   - successive harvest regrowth almost eliminated — Cottam and Nichols, 1970
   - indigenous species may reinvade — Wile et al., 1979

4. reduction in invertebrate habitat — Breck and Kitchell, 1979

5. may increase the quality of fish habitat — Wile et al., 1979

6. alterations in fish feeding and predation behavior — Breck and Kitchell, 1979; Crowder and Cooper, 1979

B. Trees

1. increase in light intensity below canopy — Clarke and Clarke, 1979

2. increased gowth of algae — Clarke and Clarke, 1979

3. increase in water temperature — Clarke and Clarke, 1979

4. conditions favour warmwater species — Clarke and Clarke, 1979

5. decreased value as winter habitat for wildlife — Clarke and Clarke, 1979

6. loss of litter source — Clarke and Clarke, 1979

7. loss of habitat for many species of mammals and birds — Clarke and Clarke, 1979

Table 4. Continued.

SYMPTOMS	REFERENCES
C. Peat	
1. may increase the local water table	Boelter and Verry, 1977
2. deterioration of water quality	Farnham, 1979
3. destruction of natural vegetation and wildlife habitat	Farnham, 1979
D. Muskrat	
1. development of a dense senescent wetland	Weller, 1978

ECOLOGICAL STRUCTURAL BREAKDOWN

STRESS: PESTS (INCLUDING BOTH EXOTIC AND NONEXOTIC SPECIES)

Marsh/Swamp

A. Vegetation (e.g., Typha glauca, Myriophyllum spicatum and Elodea canadensis)	
1. reduce the extent of open water	Clarke and Clarke, 1979
2. reduce phototrophic zone	Clarke and Clarke, 1979
3. enhance silt accumulation	Clarke and Clarke, 1979
4. increase density of vegetation	Clarke and Clarke, 1979
5. vegetation community changes	Clarke and Clarke, 1979
-shift in species composition	Clarke and Clarke, 1979
-reduction in diversity	Clarke and Clarke, 1979
-submergent community particularly threatened	Clarke and Clarke, 1979
6. reduction in habitat diversity	Clarke and Clarke, 1979

7. loss of water/cover edge habitat   Clarke and Clarke, 1979
8. reduction in food for waterfowl   Clarke and Clarke, 1979

B. Carp
  1. significant reduction in vegetation,
      primarily early in growing season   King et al., 1967; Weller, 1978
  2. some vegetation growth retarded through
      feeding and uprooting   King et al., 1967
  3. erosion of dykes and spoil banks
  4. increased turbidity   Carpenter, 1979

C. Muskrat
  1. emergent vegetation can be destroyed by   Van der Valk and Davis, 1978a;
      muskrat feeding and lodge-building   Weller, 1978

D. Beaver
  1. increased water levels on upstream side of dam   Jeglum, 1975
  2. decreased water levels on downstream side of dam   Jeglum, 1975
  3. damage to adjacent trees and shrubs   Jeglum, 1975

STRESS: FIRE   ECOLOGICAL STRUCTURAL BREAKDOWN

Marsh/Swamp
A. Physical/Chemical changes
  1. may reduce soil acidity   Smith, 1970
  2. increased release of nutrients in the ash   Smith, 1970; Vogl, 1980
  3. enhanced decomposition   Vogl, 1980
  4. reduced standing litter   Davis and Van der Valk, 1978

Table 4. Continued.

SYMPTOMS	REFERENCES
B. Vegetation changes	
1. short-term removal of rough vegetation	Smith, 1970
2. increased production of preferred wildlife foods	Smith, 1970
3. loss of habitat for some species	Weller and Spatcher, 1965
4. increase in diversity of vegetation	Clarke and Clarke, 1979
ECOLOGICAL STRUCTURAL BREAKDOWN	
STRESS: ICE DAMAGE	
Marsh	
1. development of ice foot in winter	Geis, 1979
2. with melting and ice movement in spring	
-lifting and turning over of sediments	Geis, 1979
-roots, rhizomes, etc., dislodged and exposed	Geis, 1979
-large sections of wetland edge broken off	Geis, 1979
3. loss of wetland edge habitat	Geis, 1979

[1]Source: Patterson (1984)

Table 5. A partial list of secondary stresses: associated with selective cultural stresses.[1]

Selective Cultural Stresses	Associated Secondary Stresses[2]														
	RL	CT	FI	VR	ER	SD	TU	WD	BC	NL	TL	RF	RO	RN	PE
HYDROLOGIC FLOW MODIFICATION															
Increase in water depth		*													
Decrease in water depth		*	*												
Abnormal fluctuations in depth			*	*											
Reduction in lateral flow		*		*	*	*	*	*						*	
Channelization of lateral flow				*	*	*	*	*	*				*		
WATER QUALITY DEGRADATION															
Nutrient loading (N and P)															
Toxic loading					*	*									
Thermal loading						*	*	*							
Sedimentation								*	*	*					
Turbidity		*													
ECOLOGICAL STRUCTURAL BREAKDOWN															
Bottom configuration		*			*	*			*	*	*	*			*
Biomass export		*			*	*			*	*	*	*			

Table 5. Continued.

ECOLOGICAL STRUCTURAL BREAKDOWN (Cont.)

Selective Cultural Stresses	Associated Secondary Stresses[2]														
	RL	CT	FI	VR	ER	SD	TU	WD	BC	NL	TL	RF	RO	RN	PE
Pests															
Fire						*									
Ice damage				*											

[1]Source: Patterson (1984)
[2]Key:
RL = Reduction in light penetration
CT = Change in thermal regime
FI = Increased frequency of fire
VR = Vegetation removal
ER = Erosion
SD = Sedimentation
TU = Turbidity
WD = Water depth stress
BC = Bottom configuration
NL = Nutrient loading
TL = Toxin loading
RF = Reduction in lateral flow
RD = Increase rate of surface runoff
RN = Reduced nutrients, oxygen, organics
PE = Pests

Table 6. Cultural stresses for which major programs of control have been implemented in Canadian Great Lakes shore zone water.

Stress[2]	Lake Ontario	Lake Erie	Lake St.Clair	Lake Huron	Lake Superior
Entrainment and impinge-ment of biota	*	*	*	*	*
Water level manipulation					
Exotic species					
Ice damage					
Toxin loading	*	*	*	*	*
Nutrient loading	*	*	*	*	*
Erosion and sedimentation					
Landfill					
Dredging	*	*	*	*	*
Draining and dyking					
Thermal loading	*	*	*	*	*
Hazardous spills	*	*	*	*	*
Biotic harvest	*	*	*	*	*

[1]Source: Whillans (1984)
[2]The term "stress" here has a slightly different connotation than in the text.

a setting, framework and thus a beginning for such an analysis. One aspect which has not been addressed is the control of stresses. This has been examined in a preliminary manner by Whillans (1984) with respect to Canadian shore zone fisheries.

Table 6 summarizes major ecologically oriented programs of control for 14 major stresses on the shore zone fishery. It is based on a compilation of documents produced by the International Joint Commission, Great Lakes Fishery Commission, Ontario Ministry of Natural Resources, Ontario Ministry of the Environment, regional government and other publications. These stresses for which control is indicated are not necessarily under complete control. Whillans (1984) argued that there is "little evidence of systemic ecologically motivated control of water level manipulation, exotic species, ice management, landfill and draining and dyking of wetlands." Certainly, the capability of implementing controls has been demonstrated in many small scale individual cases. A challenge remains to expand that capability to the scale of the Great Lakes.

DISCUSSION

NEEDED: A PRIMARY FOCUS ON COASTAL WATERS IN BINATIONAL GOVERNANCE OF THE GREAT LAKES

By

H. A. Regier
Institute for Environmental Studies
University of Toronto
Toronto, Ontario   M5S 1A4

Binational governance of various types and at varying degrees of formality now operates with respect to many major uses and features of the Great Lakes. Historically, these tended to be focused on events and processes that occurred at the imaginary boundary between the two nations. Vestiges of this focus may still persist, as in the implied emphasis on offshore or mid-lake phenomena in some binational programs. But ecological understanding of these lakes has grown since early studies a century ago, and it is now apparent to all that to focus studies and practices on that imaginary boundary would be a very costly way to achieve the ends of the various shared objectives on these lakes.

In many ways the offshore ecological structures and processes are dependent on natural structures and processes and on human events on land, at the land-water edge and in the near-shore area. This is where most (but not all) of the destructive abuses of humans occur, and the effects of those abuses then ramify or cascade into the offshore area. We now know that we must deal with them in the lower tributaries, in the estuaries, in the wetlands, and in the nearshore areas (including reefs, bars and shoals). I use the term "coastal waters" to encompass the habitats above.

Binational governance of the Great Lakes now relates to the following concerns:

Water quality: 1972 and 1978 Great Lakes Water Quality Agreements supervised by the International Joint Commission (IJC).

Water quantity/flows/levels: International Boards of Control of IJC, Niagara Power Treaty, Council of Great Lakes Governors and Premiers.

Fish quality and quantity: Great Lakes Fishery Commission.

Migratory birds including shore birds, raptors and waterfowl: Migratory Birds Convention.

Water-borne transportation: St. Lawrence Seaway Authority and Saint Lawrence Seaway Development Corporation, International Association of Great Lakes Ports, Great Lakes Commission.

Emergency measures for spills of hazardous materials, etc.: coordinated planning and action by Coast Guards, Canada Department of Public Works, U.S. Corps of Engineers, etc.

With respect to each of these six institutional connections central to binational governance of the Great Lakes, the lower tributaries, near-shore waters, estuaries and wetlands (or "coastal waters") are now being "administered" jointly but separately, to a large degree. For the renewable resource issues concerning fish, waterfowl and other biota, habitats on coastal waters are of crucial importance ecologically and hence practically. For water quality considerations, these areas are of key concern; all of the "areas of concern" identified pursuant to the Water Quality Agreement are in coastal waters of the Basin ecosystem. The ecological and economic impacts on coastal waters that have occurred in the past, and that are likely to occur with future proposals, are of primary importance in the making of decisions concerning new development proposals.

With respect to present and future management of coastal waters, the affected people in the immediate locale must be involved closely. But the benefits of coastal waters usually extend far beyond the immediate locale, through hydrological, ecological and cultural processes. Coastal systems modulate human impacts from land-based or inshore activities on the offshore parts of the lakes and also impacts from offshore activities on the coastline. Coastal waters generate the young of important fish and bird species that then migrate far from their natural areas. Coastal waters attract recreational interests from far away into a locality. Thus a regional financial or other subsidy for the sensitive management of coastal waters is clearly in order.

It is now time that the planning for and management of the tributaries, estuaries, wetlands and near-shore achieve better binational institutional structure and process. Two types of organizations that might be set up are a Coastal Waters Organization and a new Board under the IJC.

An intergovernmental Coastal Waters Organization might be responsible for fostering joint research, assisting in information sharing and providing management advice with respect to all aspects of these habitats. It should approach them analytically and integratively within an ecosystem approach.

A new Great Lakes Water Quality Agreement might contain a new article that focuses specifically on the coastal waters. A new IJC Board might be established to proceed with the planning and management of coastal waters, including areas of concern and connecting channels.

Such an organization is needed to preserve or conserve or rehabilitate these economically and culturally important subsystems in a more effective and efficient way than is occurring at present.

One is needed because the ecosystem approach to be effective requires that these crucial coastal waters be at the focus of attention for the comprehension and management of the whole chain of lakes.

One is needed because existing institutional arrangements for these subsystems are too fragmentary and weak, and operate only in an indirect way.

LITERATURE CITED

Adriano, D.C., A. Fulenwider, R.R. Sharitz, T.C. Ciravolo, and G.D. Hoyt. 1980. Growth and mineral nutrition of cattail (Typha) as influenced by thermal alteration. J. Environ. Qual. 9(4):649–653.

Allen, H.H. 1979. Role of wetland plants in erosion control of riparian shorelines. In: Wetland Functions and Values: The State of Our Understanding. Proc. of the National Symposium on Wetlands. P.E. Gresson, J.R. Clark, and J.E. Clark (eds.), Lake Buena Vista, pp. 403–414.

Allen, H.L. 1971. Primary productivity, chemo-organotrophy and nutritional interactions of epiphytic algae and bacteria on macrophytes of the littoral of a lake. Ecolog. Monogr. 41:97–127.

Anderson, W.L. 1948. Level ditching to improve muskrat marshes. J. Wildl. Manage. 12(2):172–176.

Andrews, C.B. 1977. The impact of the Columbia Generating Station on the surface and ground water hydrology of a river flood plain. In: Wetlands Ecology, Values and Impacts. Proc. of Waubesa Conference on Wetlands. C.B. DeWitt and E. Soloway (eds.), Madison, June 2-5, 1977, pp. 64-76.

Auclair, A.N., A. Bourhard, and J. Pajaczkowsi. 1973. Plant composition and species relations on the Huntington Marsh, Quebec. Can. J. Bot. 51:1231-1247.

Aulio, K. 1980. Accumulation of copper in fluvial sediments and yellow water lilies (Nuphar lutea) at varying distances from a metal processing plant. Bull. Environ. Contam. Toxicol. 25:713-717.

Barstow, C.J. 1971. Impact of channelization on wetland habitat in the Obion-Forked Deer Basin, Tennessee. N. Am. Wildl. Conf. 36:362-376.

Bedford, B. 1977. Seasonally displaced water temperatures as a factor affecting depletion of stored carbohydrates in Typha latifolia. In: Wetlands Ecology, Values and Impacts. Proc. of the Waubesa Conference on Wetlands, C.B. DeWitt and E. Soloway (eds.), Madison, June 2-5, 1977, pp. 83-98.

Bellrose, F.C. 1950. The relationship of muskrat populations to various marsh and aquatic plants. J. Wildl. Manage. 14:299-315.

Birch, P.B., R.S. Barnes, and D.E. Spyridakis. 1980. Recent sedimentation and its relationship with primary productivity of 4 western Washington lakes. Limnol. and Ocean. 25:240-247.

Boelter, D.H. and E.S. Verry. 1977. Peatland and Water in the Northern Lake States. U.S. Dept. Agric., For. Serv., Gen. Tech. Rep. NC-31, 22 pp.

Boto, K.G. and W.H. Patrick, Jr. 1979. Role of wetlands in the removal of suspended sediments. In: Wetland Functions and Values: The State of Our Understanding. Proc. of the National Symposium on Wetlands. P.E. Gresson, J.R. Clark, and J.E. Clark (eds.), Lake Buena Vista, pp. 479-489.

Bott, T.L. 1975. Bacterial growth rates and temperature optima in a stream with fluctuations in thermal regime. Limnol. and Ocean. 20(2):191-197.

Boyd, C.E. 1978. Chemical composition of wetland plants. In: Freshwater Wetlands. Ecological Processes and Management

Potential. R.E. Good, D.F. Whigham, R.L. Simpson (eds.), Academic Press, New York, pp. 155-167.

Boyt, F.L., S.E. Bayley, and J. Zoltek, Jr. 1977. Removal of nutrients from treated municipal wastewater by wetland vegetation. J. Wat. Poll. Cont. Fed. 49(5):789-799.

Brady, R.F. 1979. Agricultural Land Drainage in Ontario. Dept. of Geography, Brock University, St. Catharines, Ontario. Unpublished, 17 pp.

Breck, J.E. and J.F. Kitchell. 1979. Effects of macrophyte harvesting on simulated predator-prey interactions. In: Aquatic Plants, Lake Management and Ecosystem Consequences of Lake Harvesting. J.E. Breck, R.T. Prentki, O.L. Loucks (eds.), University of Wisconsin-Madison, pp. 211-228.

Cairns, J. and K. Dickson. 1977. Recovery of streams from spills of hazardous materials. In: Recovery and Restoration of Damaged Ecosystems. J. Cairns, Jr., K.L. Dickson, and E.E. Herricks (eds.), University Press of Virginia, pp. 24-42.

Carpenter, S.R. 1979. The invasion and decline of Myriophyllum spicatum in a eutrophic Wisconsin lake. In: Aquatic Plants, Lake Management and Ecosystem Consequences of Lake Harvesting. J.E. Breck, R.T. Prentki, O.L. Loucks (eds.), University of Wisconsin-Madison, pp. 43-50.

Christy, E.J. and R.R. Sharitz. 1980. Characteristics of 3 populations of a swamp annual under different temperature regimes. Ecology. 61(3):454-460.

Clarke, J. and J. Clarke. 1979. Scientists' Report. The National Symposium on Wetlands. Florida, Nov. 6-9, 1978, National Wetlands Technical Council, 129 pp.

Cook, A.H. and C.F. Powers. 1958. Early biochemical changes in the soils and waters of artificially created marshes in New York. N.Y. Fish and Game J. 5:9-65.

Cooke, G.D. 1980. Lake level drawdown as a macrophyte control technique. Water Res. Bull. 16(2):317-322.

Cooley, J.M. 1971. The effect of temperature on the development of resting eggs of Diatomus oregonensis. Limnol. and Ocean. 16(6):921-926.

Cottam, G. and S.A. Nichols. 1970. Changes in the Water Environment Resulting from Aquatic Plant Control. University of Wisconsin Water Research Center Tech. Rep. OWRRB–019–W13, 27 pp.

Crowder, L.B. and W.E. Cooper. 1979. The effects of macrophyte removal on the feeding efficiency and growth of sunfishes: Evidence from pond studies. In: Aquatic Plants, Lake Management and Ecosystem Consequences of Lake Harvesting. J.E. Breck, R.T. Prentki, O.L. Loucks (eds.), University of Wisconsin-Madison, pp. 251-268.

Cushing, Jr., C.E. and J.M. Thomas. 1980. Cu and Zn kinetics in Myriophyllum heterophyllum michx. and Potamogeton richardsonii (Ar. Bene) Rydb. Ecology. 61(6):1321-1326.

Darnell, R.M., W.E. Pequegnat, B.M. James, F.J. Benson, and R.A. Defenbaugh. 1977. Impacts of Construction Activities in Wetlands of the United States. U.S. Environmental Protection Agency, Ecol. Res. Ser. EPA-600/3-76-045, 393 pp.

Davis, C.B. and A.G. van der Valk. 1978. Litter decomposition in prairie glacial marshes. In: Freshwater Wetlands. Ecological Processes and Management Potential. R.E. Good, D.F. Whigham, R.L. Simpson (eds.), Academic Press, New York, pp. 99-113.

Davis, S.M. and L.A. Harris. 1978. Marsh plant production and phosphorus flux in everglades conservation area 2. In: Environmental Quality through Wetlands Utilization. M.A. Drew (ed.), Tallahassee, Florida, pp. 105-131.

Dolan, T.J., S.E. Bayley, J. Zolteck, Jr., and A. Hermann. 1978. The Clermont Project: Renovation of treated effluent by a fresh water marsh. In: Environmental Quality through Wetlands Utilization. M.A. Drew (ed.), Tallahassee, Florida, pp. 132-152.

Erickson, C. and D.C. Mortimer. 1975. Mercury uptake in rooted higher aquatic plants; laboratory studies. Verh. Internat. Verein. Limnol. 19:2087-2093.

Farnham, R.S. 1979. Wetlands as energy sources. In: Wetland Functions and Values: The State of Our Understanding. Proc. of the National Symposium on Wetlands. P.E. Gresson, J.R. Clark, and J.E. Clark (eds.), Lake Buena Vista, pp. 661-672.

Found, W.C., A.R. Hill, and E.S. Spence. 1974. Economic and Environmental Impacts of Land Drainage in Ontario. York University, Geographical Monographs #6, 175 pp.

Francis, G.R., A.P. Grima, H.A. Regier, and T.H. Whillans. 1985. A Prospectus for the Management of the Long Point Ecosystem. Great Lakes Fishery Commission, Tech. Rep. (in press), 202 pp.

Francis, G.R., J.J. Magnuson, H.A. Regier, and D.R. Talhelm (eds.). 1979. Rehabilitating Great Lakes Ecosystems. Great Lakes Fishery Commission, Tech. Rep. No. 37, 99 pp.

Friend, M., G.E. Cummings, and J.S. Morse. 1964. Effect of changes in winter water levels on muskrat weights and harvest at the Montezuma National Wildlife Refuge. N.Y. Fish and Game J. 11(2):125-131.

Geis, J.W. 1979. Shoreline processes affecting the distribution of wetland habitat. Trans. N. Am. Wildl. Conf. 44:529-542.

Glooschenko, W.A. and J.A. Capobianco. 1978. Metal content of sphagnum mosses from two northern Canadian bog ecosystems. Water, Air and Soil Pollution. 10:215-220.

Gorham, E. and D.L. Tilton. 1978. The mineral content of Sphagnum fuscum as effected by human settlement. Can. J. Bot. 56:2755-2759.

Gosselink, J.G. and R.E. Turner. 1978. The role of hydrology in freshwater wetland ecosystems. In: Freshwater Wetlands. Ecological Processes and Management Potential. R.E. Good, D.F. Whigham, R.L. Simpson (eds.), Academic Press, New York, pp. 63-78.

Great Lakes Basin Commission/Erosion and Sedimentation Work Group. 1975. Erosion and Sedimentation. Appendix 18 to the Great Lakes Basin Framework Study, 129 pp.

Groen, C.L. and J.C. Schmalbach. 1978. The sport fishery of the unchannelized and channelized Middle Missouri River. Trans. Am. Fish. Soc. 107(3):412-418.

Hamley, J.M. and N.G. MacLean. 1979. Impacts of Nanticoke industrial development. Contact. 11(1):81-115.

Harms, W.R., H.T. Schreuder, D.D. Hook, C.L. Brown, and F.W. Shropshire. 1980. The effects of flooding on the swamp forest in Lake Ocklawaha, Florida. Ecology. 61(6):1412-1421.

Harris, H.J., D.R. Talhelm, J.J. Magnuson, and A.M. Forbes. (eds). 1982. Green Bay in the Future--A Rehabilitative Prospectus.

Great Lakes Fishery Commission, Tech. Rep. No. 38, 59 pp.

Harris, S.W. and W.H. Marshall. 1963. Ecology of water-level manipulations on a northern marsh. Ecology. 44:331-343.

Hartland-Rowe, R. and P.B. Wright. 1975. Effects of sewage effluent on a swampland stream. Verh. Internat. Verein. Limnol. 19(2):1575-1583.

Herdendorf, C.E. and S.M. Hartley (eds.). 1979. A Summary of the Knowledge of the Fish and Wildlife Resources of the Coastal Wetlands of the Great Lakes of the United States. Volume 1: Overview, Ohio State University and Indiana University. Draft, 468 pp.

Hynes, H.B.N. 1974. The Biology of Polluted Waters. University of Toronto Press, 202 pp.

International Joint Commission. 1980. Lake Erie Water Level Study. Appendix F. Environmental Effects. Draft manuscript.

Jaworski, E., C.N. Raphael, P.J. Mansfield, and B.B. Williamson. 1979. Impact of Great Lakes water level fluctuations on coastal wetlands. In: A Summary of the Knowledge of the Fish and Wildlife Resources of the Coastal Wetlands of the Great Lakes of the United States. Herdendorf, C.E. and S.M. Hartley (eds.), Ohio State University and Indiana University. Draft, pp. 103-297.

Jeglum, J.K. 1975. Vegetation - habitat changes caused by damming a peat drainageway in northern Ontario. Can. Field Natur. 89:400-412.

Johnson, D.L. and W.S. Snyder. 1984. Ohio State University, personal communication.

Kadlec, R.H. and J.A. Kadlec. 1979. Wetlands and water quality. In: Wetland Functions and Values: The State of Our Understanding. Proc. of the National Symposium on Wetlands. P.E. Gresson, J.R. Clark, and J.E. Clark (eds.), Lake Buena Vista, pp. 436-456.

Kadlec, J.A. 1962. The effects of a drawdown on a waterfowl impoundment. Ecology. 43:267-281.

King, D.R. and G.S. Hunt. 1967. Effect of carp on vegetation in a Lake Erie shooting club. J. Wildl. Manage. 31(1):181-188.

Klopatek, J.M. 1978. Nutrient dynamics of freshwater riverine marshes and the role of emergent macrophytes. In: Freshwater Wetlands. Ecological Processes and Management Potential. R.E. Good, D.F. Whigham, R.L. Simpson (eds.), Academic Press, New York, pp. 217-243.

Krenkel, P.A. and F. Parker (eds.). 1969. Biological Aspects of Thermal Pollution. Vanderbilt University Press.

Lake Erie Regulation Study. 1981. Report to the International Joint Commission by the International Lake Erie Regulation Study Board, 230 pp.

Larson, G. 1981. Effects of wetland drainage on surface runoff. In: Selected Proceedings of the Midwest Conference on Wetland Values and Management. B. Richardson (ed.), St. Paul, Minnesota, pp. 117-120.

Lathwell, D.J., H.F. Mulligan, and D.R. Bouldin. 1969. Chemical properties, physical properties and plant growth in twenty artificial wildlife marshes. N.Y. Fish and Game J. 16(2):158-183.

Lee, G.F., E. Bentley, and R. Amundson. 1975. Effects of marshes on water quality. In: Coupling of Land and Water Systems. A.D. Hasler (ed.), Springer-Verlag, New York, pp. 105-127.

Loucks, O.L. 1981. The littoral zone as a wetland: Its contribution to water quality. In: Selected Proceedings of the Midwest Conference on Wetland Values and Management. B. Richardson (ed.), St. Paul, Minnesota, pp. 125-137.

McCullough, G. 1981. Wetland losses in Lake St. Clair and Lake Ontario. In: Proc. of the Ontario Wetlands Conference. A. Champagne (ed.), Toronto, Ontario, pp. 81-89.

McLeese, R.L. and E.P. Whiteside. 1977. Ecological effects of highway construction upon Michigan woodlots and wetlands: Soil relationships. J. Environ. Qual. (4):467-471.

McNaughton, S.J., T.C. Folsom, T. Lee, F. Park, C. Price, D. Roeder, J. Schmitz, and C. Stockwell. 1974. Heavy metal tolerance in Typha latifolia without the evolution of tolerant races. Ecology. 55(5):1163-1165.

Meeks, R.L. 1969. The effect of drawdown date on wetland plant succession. J. Wildl. Manage. 33(4):817-821.

Millar, J.B. 1973. Vegetation changes in shallow marsh wetlands under improving moisture regime. Can. J. Bot. 51:1443–1457.

Nicholls, K.H. and H.R. MacCrimmon. 1974. Nutrients in subsurface and runoff waters of the Holland Marsh, Ontario. J. Environ. Qual. 3(1):31–35.

Patrick, R. 1978. Some impacts of channelization on riverine systems. In: Environmental Quality through Wetlands Utilization. Proc. of a Symposium on Freshwater Wetlands. M.A. Drew (ed.), Tallahassee, Florida, Feb. 28–March 2, 1979.

Patterson, N.J. 1984. An Approach to Wetland Evaluation and Assessing Cultural Stresses on Freshwater Wetlands in Southern Ontario. MSc Thesis, Institute for Environmental Studies, University of Toronto.

Prentki, R.T., T.D. Gustafson, and M.S. Adams. 1978. Nutrient movements in lakeshore marshes. In: Freshwater Wetlands. Ecological Processes and Management Potential. R.E. Good, D.F. Whigham, R.L. Simpson (eds.), Academic Press, New York, pp. 169–194.

Reddy, C.N. and W.H. Patrick, Jr. 1977. Effect of redox potential and pH on the uptake of cadmium and lead by rice plants. J. Environ. Qual. 63(3):259–262.

Reed, D.M., J.H. Riemer, and J.A. Schwarzmeier. 1977. Some observations on the relationship of floodplain siltation to reed canary grass abundance. In: Wetlands Ecology, Values and Impacts. Proc. of the Waubesa Conference on Wetlands, C.B. DeWitt and E. Soloway (eds.), Madison, June 2–5, 1977, pp. 99–107.

Richardson, C.J., D.T. Tilton, J.A. Kadlec, J.P.M. Chamie, and W. A. Wentz. 1978. Nutrient dynamics of northern wetland ecosystems. In: Freshwater Wetlands. Ecological Processes and Management Potential. R.E. Good, D.F. Whigham, R.L. Simpson (eds.), Academic Press, New York, pp. 217–242.

Sharitz, R.R., J.E. Irwin, and E.J. Christy. 1979. Vegetation of swamps receiving reactor effluents. Oikos. 25:7–13.

Sloey, W.E., F.L. Sprangler, and C.W. Fetter, Jr. 1978. Management of freshwater wetlands for nutrient assimilation. In: Freshwater Wetlands. Ecological Processes and Management Potential. R.E. Good, D.F. Whigham, R.L. Simpson (eds.), Academic Press, New York, pp. 321–340.

Smith, E.R. 1970. Evaluation of a leveed Louisiana marsh. Trans. North Amer. Wildl. Conf. 35:265-275.

Stone, J.H., L.M. Bahr, Jr., and J.W. Day, Jr. 1978. Effects of canals on freshwater marshes in coastal Louisiana and implications for management. In: Freshwater Wetlands. Ecological Processes and Management Potential. R.E. Good, D.F. Whigham, R.L. Simpson (eds.), Academic Press, New York, pp. 299-320.

Van der Valk, A.G. and L.C. Bliss. 1971. Hydrarch succession and net primary production of oxbow lakes in central Alberta. Can. J. Bot. 49:1177-1199.

Van der Valk, A.G. and C.B. Davis. 1978a. Primary production of prairie glacial marshes. In: Freshwater Wetlands. Ecological Processes and Management Potential. R.E. Good, D.F. Whigham, R.L. Simpson (eds.), Academic Press, New York, pp. 21-37.

Van der Valk, A.G. and C.B. Davis. 1978b. The role of seed banks in the vegetation dynamics of prairie glacial marshes. Ecology. 59(2):322-335.

Vogl, R.J. 1980. The ecological factors that produce perturbation - dependent ecosystems. In: The Recovery Process in Damaged Ecosystems. J. Cairns (ed.), Ann Arbor Science Publishers, Inc., Ann Arbor, Michigan, pp. 63-95.

Weller, M.W. 1978. Management of freshwater marshes for wildlife. In: Freshwater Wetlands. Ecological Processes and Management Potential. R.E. Good, D.F. Whigham, R.L. Simpson (eds.), Academic Press, New York, pp. 267-284.

Weller, M.W. and G.S. Spatcher. 1965. Role of Habitat in the Distribution and Abundance of Marsh Birds. Special Report #43, Iowa State University of Science and Technology, Ames, Iowa, April, 1965.

Welsh, P. and P. Denny. 1979. Waterplants and the recycling of heavy metals in an English lake. In: Trace Substances in Environmental Health-X. D.D. Hemphill (ed.), University of Missouri, Columbia, pp. 217-223.

Whigham, D.F. and R.L. Simpson. 1976. The potential use of freshwater tidal marshes in the management of water quality in the Delaware River. In: Biological Control of Water Pollution. J. Tourbier and R.W. Pierson (eds.), University of Pennsylvania Press, Philadelphia, PA, pp. 173-186.

Whillans, T.H. 1984. Fish habitat alterations and present management practices along the Canadian shore zone of the Laurentian Great Lakes. In: Proc. of the Non-Salmonid Rehabilitation Workshop. October 1-3, 1984. Ontario Ministry of Natural Resources (in press).

Whillans, T.H. 1982. Changes in marsh area along the Canadian shore of Lake Ontario. J. Great Lakes Res. 8(3):570-577.

Whillans, T.H. 1980. Feasibility of Rehabilitating the Shore Zone Fishery of Lake Ontario. Resource Document #11 for Lake Ontario Tactical Fisheries Plan, Ontario Ministry of Natural Resources, 188 pp.

Wile, I. 1975. Lake restoration through mechanical harvesting of aquatic vegetation. Verh. Internat. Verein. Limnol. 19:660-671.

Wile, I., G. Hitchin, and G. Beggs. 1979. Impact of mechanical harvesting on Chemung Lake. In: Aquatic Plants, Lake Management and Ecosystem Consequences of Lake Harvesting. J.E. Breck, R.T. Prentki, O.L. Loucks (eds.), University of Wisconsin-Madison, pp. 145-159.

CHAPTER 15

CONTROL OF CATTAIL AND BULRUSH
BY CUTTING AND FLOODING

Richard M. Kaminski[1]
Ducks Unlimited Canada
1190 Waverley Street
Winnipeg, Manitoba R3T 2E2

Henry R. Murkin
Delta Waterfowl and Wetland Research Station
RR 1
Portage la Prairie, Manitoba R1N 3A1

Christopher E. Smith
Ducks Unlimited Canada
Box 1799
The Pas, Manitoba R9A 1L5

INTRODUCTION

Managers of wetland habitats frequently encounter dense, unbroken stands of emergent vegetation that are unattractive to waterfowl and marsh birds. Interspersion of emergent vegetation and open water can be enhanced using methods that Weller (1981) categorized as "natural" (e.g., water-level manipulation, fire) or "artificial" (e.g., ditching, mowing). Cutting emergent vegetation represents an artificial technique not widely used, despite its effectiveness in inhibiting revegetation and enhancing emergent vegetation-water interspersion.

---

[1]Present address: Department of Wildlife and Fisheries, Mississippi State University, P.O. Drawer LW, Mississippi State, MS 39762.

Although several studies have documented the effect of cutting during the growing season on regeneration of cattail (Typha spp.) (Nelson and Dietz, 1966; Weller, 1975; Beule, 1979; Sale and Wetzel, 1983), the effect of cutting during or after the growing season on regeneration of bulrush (Scirpus spp.) has not been reported to our knowledge. Moreover, operation of heavy cutting machinery in wetlands is often difficult during the growing season due to soft substrates. At northern latitudes, cattail has been cut and disked over frozen substrates (Sanderson and Bellrose, 1969; Weller, 1975; Murkin and Ward, 1980), thus increasing the practicality of the technique. However, rates of cutting emergent vegetation over frozen substrates using different farm implements have not been published. The objectives of our study were to: (1) evaluate the combined effect of post-growing season cutting and spring-summer flooding as a technique to control regeneration of three common emergent species (common cattail, T. latifolia; tule bulrush, S. acutus; and softstem bulrush, S. validus), and (2) quantify the relative efficiency of different implements used in cutting emergent vegetation over frozen substrates.

STUDY AREA AND METHODS

Our study was conducted on several marshes in western Canada (Table 1). Within each marsh, monotypic stands of the target plant species were selected for cutting. Cutting procedures, implements, and size and configuration of cut areas differed among study areas and are described below for each area. Regeneration of cattail and bulrush was investigated at Milligan Creek and Kapakoskasewakak Lake, respectively. Only the relative efficiency of the specific cutting implements was determined at Pelican Lake North and Delta Marsh.

Milligan Creek

Cattail was strip-cut (ca. 3 m wide) over ice in late November-early December, 1979, amidst snow depths ranging between 8-46 cm in the study area. Two implement combinations were used: (1) a 50-hp tractor equipped with a 1.5 m wide rotary mower, and (2) a self-propelled 1.7 m wide swather. To reduce stress imposed upon the swather by the cut leaves and stems, the collecting table and canvas belts of the swather were removed which caused severed cattail to fall directly to the ice.

Mowed and swathed areas at Milligan Creek could not be clearly distinguished in mid-May, 1980, under flooded conditions. Therefore, mowed and swathed areas were merely

Table 1.   Target plant species and locations of four study areas.

Plant Species	Study Area	Location
Common cattail	Milligan Creek	Central Saskatchewan (51° 52' N, 103° 51' W)
	Delta Marsh	South central Manitoba (50° 11' N, 98° 19' W)
Tule bulrush	Pelican Lake North	East central Saskatchewan (52° 47' N, 105° 43' W)
Tule and softstem bulrush	Kapakoskasewakak Lake	Northwestern Manitoba (53° 39' N, 100° 8' W)

designated as cut areas for subsequent analysis of cattail regeneration data. To establish sample areas for evaluating cattail regeneration in response to cutting and flooding, 90 square quadrats (1 m^2 each) were randomly located and marked in mid-May, 1980, in both cut and uncut (i.e., control) areas. Three variables were measured at that time within
each quadrat in the cut areas: (1) water depth (cm) from surface to substrate, (2) water depth (cm) above cattail stubble of maximum length, and (3) maximum stubble length (cm). In mid-August, 1980 and 1981, water depth from surface to substrate was again measured at all relocated sample cut sites. In addition, three vegetative variables were measured within quadrats at all relocated cut and uncut sample sites: (1) number of new cattail shoots/m^2, (2) number of new flowered shoots/m^2, and (3) maximum leaf length (cm).

Kapakoskasewakak Lake (KL)

In early August, 1981, while KL was completely drained, 24 circular plots (ca. 7 m^2 each) were randomly located within monotypic stands of tule and softstem bulrush. Using machetes, vegetation was handcut at ground level (stubble length ca. 5 cm) within each circular plot).

In mid-August, 1982, the density of new and flowered shoots of tule and softstem bulrush was determined within square quadrats (0.25 m^2) at all relocated sample cut sites and at 24 randomly located uncut sites within the same stands of both bulrush species. Water depth was measured in May and August, 1982, at 188 random locations throughout the stands of both bulrush species that contained sample sites.

Pelican Lake North

A 50-hp tractor and rotary mower (1.5 m) was used to cut sinuous strips and circular openings over ice within stands of tule bulrush. The manipulation was performed in mid-January, 1982, amidst approximately 25 cm of snow.

Delta Marsh

Murkin and Ward (1980) detailed their procedures of cutting cattail in the Delta Marsh. Briefly, however, they cut cattail at ground level during April, 1978, using a tractor-drawn 7-blade disc (1.3 m wide), traveling over frozen earth in water depths between 1-45 cm.

RESULTS AND DISCUSSION

Cattail Regeneration

Mean densities of new and flowered shoots of cattail and mean maximum leaf length of cattail were less (P < 0.001, t-test) on cut than uncut areas at Milligan Creek in 1980 and 1981 (Table 2). Autumn cutting of cattail over ice and subsequent flooding during all or part of both growing seasons negatively influenced cattail regeneration. However, the effect of cutting and flooding on decreased regeneration of cattail in 1981 may have been confounded with drought conditions that year. Average water levels in mid-May, 1981, at 85 sample sites was 8.6 cm (SE = 0.39 cm) compared to 34.8 cm (SE = 0.35 cm) at the same sites in 1980. Surface water was nonexistent at all 85 sites in mid-August, 1981, whereas only 8 of 90 (9%) sites lacked surface water in August, 1980.

We initially selected five variables (D_1 = May water depth, D_2 = August water depth, D_3 = decrease in water depth between May and August, D_4 = water depth above cattail stubble of maximum length, and SL = maximum stubble length) to test for relationships between these variables and 1980 new shoot density (SD) using stepwise multiple regression. Some values of SD were zero; therefore, all values of SD were transformed (SD + 0.5) prior to regression analysis (Zar, 1974). D_1 and D_2 were correlated (r = 0.86, P < 0.001, N = 90). Thus, we excluded D_2 from the regression analysis, because our goal was to calculate a value for D_1 that corresponded to zero cattail regrowth, if a significant

Table 2. Floristic responses of common cattail in 1980 and 1981 on cut and uncut areas at Milligan Creek in Saskatchewan.

Variable	Year	Cut			Uncut		
		$\underline{N}$	$\bar{x}$	SE	$\underline{N}$	$\bar{x}$	SE
$\underline{N}$ shoots/m^2	1980	90	7.4	0.9	90	29.8	0.7
	1981	85	10.0	0.8	85	20.2	0.7
$\underline{N}$ flowered shoots/m^2	1980	76	0.2	0.1	90	5.6	0.4
	1981	77	2.2	0.4	85	6.1	0.4
Max. leaf length (cm)	1980	76	139.1	3.8	90	163.8	1.3
	1981	77	114.8	1.3	85	131.3	2.0

relationship ($\underline{P} \leq 0.05$) existed between SD and ≥ 1 explanatory variables.

Variation in SD was related ($\underline{R}^2$ = 0.41, $0.001 < \underline{P} < 0.05$, $\underline{N}$ = 90) to three of the four explanatory variables. D_1 exerted the most influence on variation in SD, as indicated by its regression coefficient in the following model:

$$\sqrt{SD + 0.5} = 6.43 - 0.26(D_1) + 0.15(D_3) + 0.04(SL)$$

To calculate the estimate of D_1 that corresponded to zero cattail regrowth, the above equation was solved setting SD = 0, D_3 = 19 cm, and SL = 17 cm. The value for D_3 was a regional long-term mean estimate for net evaporation (i.e., gross evaporation minus precipitation) for the period May-August, 1911-79, for Yorkton, Saskatchewan; the nearest weather station (140 km) to Milligan Creek. The value for SL was the 1980 $\bar{x}$ value from Milligan Creek. The solving process resulted in D_1 = 36 cm. Murkin and Ward (1980) reported an April water depth of 26 cm corresponded to no regrowth of cattail based on regressing new shoot density against April water depth. A factor possibly explaining the difference between our estimate and theirs is differential net evaporation rates between Milligan Creek and Delta Marsh. For example, the long-term May-August average net evaporation rates for Yorktown, Saskatchewan, and Winnipeg, Manitoba, are 19 and 10 cm, respectively. Other factors also could have influenced the difference,

Table 3. Floristic repsonses of tule and softstem bulrush
in 1982 on cut and uncut areas at Kapakoskasewakak Lake
in Manitoba.

Species	Variable	Cut			Uncut		
		N	x̄	SE	N	x̄	SE
Tule bulrush	N shoots/0.25 m²	22	3.4	0.7	24	46.0	2.8
	N flowered shoots/0.25 m²	16	1.0	0.3	24	29.9	2.5
Softstem bulrush	N shoots/0.25 m²	24	21.4	3.1	24	71.8	6.5
	N flowered shoots/0.25 m²	23	0.5	0.2	24	5.4	1.1

including ecotypic variation in cattail stocks between
areas, longer stubble at Milligan Creek than Delta Marsh,
and microclimatic differences between the study areas
and the nearest weather stations.

Bulrush Regeneration

Mean density of new and flowered shoots of tule
and softstem bulrush was less ($P < 0.005$, t-test) on cut
than uncut areas in KL in 1982 (Table 3). Mean water
depth at 188 sample sites was 62 cm (SE = 0.2 cm) in May,
1982, and 38 cm (SE = 0.2 cm) in August, 1982, indicating
stubble of both bulrush species was inundated throughout
the 1982 growing season. We conclude that cutting and
subsequent flooding inhibited regrowth and flowering of
tule and softstem bulrush.

Decreased Plant Regeneration by Cutting and Flooding

Murkin and Ward (1980) and Sale and Wetzel (1983)
discussed the possible mechanism by which cutting and
subsequent flooding of cattail stubble results in its
decreased regeneration. They explained that water over
stubble eliminates diffusion of atmospheric oxygen to
the rhizome. Hence, the only oxygen available to the
plant is that contained within the submerged aerenchyma.
This source of oxygen is rapidly depleted, causing anaerobic
respiration and eventually death if new shoots fail to
reach the atmosphere to supply oxygen to the rhizome.

We believe the same principles apply to tule and softstem bulrush.

Relative Efficiencies of Implements

Although the different implements used in this study were operated by different persons under different habitat conditions, we determined the implements' cutting rates to assess the relative efficiency of each for cutting dense emergent vegetation. Of the implements used, the tractor-drawn rotary mower was most efficient (Table 4). The average cutting rate of the tractor-mower combinations was two and three times greater than the rate of the swather and tractor-drawn disk, respectively. Moreover, no breakdowns occurred with the tractor-drawn rotary mowers, but the swather frequently failed during our cutting operations. Nelson and Dietz (1966) similarly reported that a tractor-drawn rotary mower was three times more efficient than tilling or disking.

MANAGEMENT IMPLICATIONS

Post-growing season cutting of cattail and bulrush with subsequent flooding during the growing season appears to be an effective control method for these and possibly other emergent plant species. The technique also enhances emergent vegetation-water interspersion in marshes and creates open-water areas that attract waterfowl and other marsh birds. An interspersed, 50:50 coverage of emergent

Table 4. Cutting rates of different implements used to cut tule bulrush or cattail over frozen substrates on three areas in western Canada.

Implement(s)	Cutting Rate (ha/hr)	Plant Species	Study Area
Tractor and rotary mower	0.59	Tule bulrush	Pelican Lake North
	0.41	Cattail	Milligan Creek
Swather	0.26	Cattail	Milligan Creek
Tractor and disc	0.19	Cattail	Delta Marsh

vegetation and open water should be most attractive to these birds (Weller and Spatcher, 1965; Weller and Fredrickson, 1974; Kaminski and Prince, 1981; Murkin et al., 1982). Waterbird use of cut areas may be further enhanced by cutting interconnections between clearcut areas (Weller, 1975). Additionally, the litter resulting from cutting operations forms a substrate for microbes that can be exploited by aquatic invertebrates and ultimately by birds. Floating litter also serves as temporary loafing and nesting sites for certain species (e.g., Anas platyrhynchos, Branta canadensis, Fulica americana), as evidenced by our observations and those other others (e.g., Beule, 1979; Krapu et al., 1979). If managers contemplate using the technique to improve wetlands for breeding waterfowl, an important consideration is the close availability of suitable nesting and brood-rearing habitats.

Responses by waterbirds to openings of different size and configuration have not been published. However, Kaminski and Prince (1981) suggested that mowed sinuous strips may be as attractive to waterbirds as clear-cut openings, but this suggestion should be field tested.

Snow and ice hinder cutting efficiency and can cause equipment failures. If substrates will support machinery, mowing operations can be initiated soon after the growing season. For example, in drawdown revegetated basins, post-growing season mowing of emergent vegetation may be useful to create openings for use by migrant and wintering waterbirds. This approach could eliminate delays in avian use that might occur before natural openings form in response to flooding, herbivory, and natural toppling of vegetation.

Emergent vegetation should be mowed as close as possible to the basin substrate to facilitate inundation of the stubble during the growing season. Success in limiting regrowth of emergents following cutting depends on the duration of stubble inundation. Precipitation and evapotranspiration vary temporally and spatially. However, a generally applicable recommendation is to confine cutting operations to basin elevations where spring water depths above average stubble length will at least equal the regional long-term net evaporation rate for spring through mid-summer. This strategy should keep stubble inundated long enough to exhaust oxygen and carbohydrate reserves in the rhizome that are important for initial growth in spring (Linde et al., 1976).

We and Nelson and Dietz (1966) found that a tractor-drawn rotary mower was most efficient for cutting dead and live emergent vegetation, respectively. However,

managers should determine the efficiencies of different
implements relative to current costs of implement purchase
or rental to forecast economic benefits. Based on our
experiences in mowing different emergent species (e.g.,
Typha spp., Scirpus spp., Phragmites communis, Scolochloa
festucacea, Carex atherodes), we recommend using a tractor
with $\geq$ 50 horsepower and a rotary mower for cutting live
or dead emergent vegetation. Cutting efficiency and dispersal
of clipped plants can be increased by removing the steel-
chain curtain attached to the rear of most rotary mowers.
Nelson and Dietz (1966) suggested other equipment improvements
including half-tracks or front and rear dual wheels to
improve support over soft substrates, power steering when
dual front wheels are employed, and a radiator screen
to prevent engine over-heating.

SUMMARY

In several marshes in western Canada, we evaluated
regeneration of common cattail, tule bulrush, and softstem
bulrush following post-growing season cutting and subsequent
spring-summer flooding as a method of control. Cutting
along with inundation of the stubble significantly decreased
total shoot density (50-93%) and flowering shoot density
(64-97%) in all three emergent species, suggesting the
technique is useful for control of emergent vegetation
and enhancement of emergent vegetation-water interspersion
for increased use of overgrown marshes by waterfowl and
marsh birds. Of several farm implements used to cut cattail
or bulrush, a tractor-drawn rotary mower was most efficient.

ACKNOWLEDGMENTS

Ducks Unlimited Canada funded this study and shared
publication costs with the Delta Waterfowl and Wetlands
Research Station, Canada. Special thanks are extended
to the staff of Ducks Unlimited Canada who assisted with
this project. D. H. Arner, S. R. Guinn, G. A. Hurst,
H. A. Jacobson, J. W. Nelson, and C. J. Newling provided
constructive reviews of the manuscript.

LITERATURE CITED

Buele, J.D. 1979. Control and management of cattails
in southeastern Wisconsin wetlands. Wis. Dept. Nat.
Resour. Tech. Bull. 112, 39 pp.

Kaminski, R.M. and H.H. Prince. 1981. Dabbling duck
and aquatic macroinvertebrate responses to manipulated
wetland habitat. J. Wildl. Manage. 45:1-15.

Krapu, G.L., L.G. Talent, and T.J. Dwyer. 1976. Marsh nesting by mallards. Wildl. Soc. Bull. 7:104–110.

Linde, A.F., T. Janisch, and D. Smith. 1976. Cattail, the significance of its growth, phenology and carbohydrate storage to its control and management. Wis. Dept. Nat. Resour. Tech. Bull. 94, 26 pp.

Murkin, H.R. and P. Ward. 1980. Early spring cutting to control cattail in a northern marsh. Wildl. Soc. Bull. 8:254–256.

Murkin, H.R., R.M. Kaminski, and R.D. Titman. 1982. Responses by dabbling ducks and aquatic invertebrates to an experimentally manipulated cattail marsh. Can. J. Zool. 60:2324–2332.

Nelson, N.F. and R.H. Dietz. 1966. Cattail control methods in Utah. Utah Dept. Fish Game Publ. 66–2, 31 pp.

Sale, P.J.M. and R.G. Wetzel. 1983. Growth and metabolism of Typha species in relation to cutting treatments. Aq. Bot. 15:321–334.

Sanderson, G.C. and F.C. Bellrose. 1969. Wildlife habitat management of wetlands. Ann. Acad. Brasil. Cienc. 41:155–204.

Weller, M.W. 1975. Studies of cattail in relation to management for marsh wildlife. Iowa State J. Sci. 49:383–412.

Weller, M.W. 1981. Freshwater marshes, ecology and wildlife management. Univ. of Minnesota Press, Minneapolis, 146 pp.

Weller, M.W. and L.H. Fredrickson. 1974. Avian ecology of a managed glacial marsh. Living Bird 12:269–291.

Weller, M.W. and C.E. Spatcher. 1965. Role of habitat in the distribution and abundance of marsh birds. Iowa State Univ., Agric. and Econ. Exp. Stn. Spec. Rep. 43, 31 pp.

Zar, J.H. 1984. Biostatistical Analysis. Prentice-Hall, Inc., Englewood Cliffs, NJ, 620 pp.

CHAPTER 16

MARSH MANAGEMENT BY WATER LEVEL MANIPULATION
OR OTHER NATURAL TECHNIQUES:
A COMMUNITY APPROACH

John P. Ball
Department of Zoology
University of Guelph
Guelph, Ontario   N1G 2W1
and
Delta Waterfowl and Wetlands Research Station
RR1 Portage la Prairie, Manitoba   R1N 3A1

INTRODUCTION

As a result of the loss of wetland habitats, many
public interest groups, conservation agencies and professional
organizations agree on the need to preserve wetlands.
Beyond this agreement, however, the diverse value systems
of these groups lead to disagreement as to what should
be done with these marshes. Some groups advocate simple
preservation of wetlands in their existing state, while
others wish to manage wetlands for the production of certain
species or taxa. Some of these disagreements may be unavoidable,
but perhaps a single-species philosophy of management
has exacerbated these differences of opinion. A community
or multi-species approach to wetland management, however,
may more likely satisfy the aims of these various interest
groups.

In this paper, I will discuss such an approach to
wetland management and will show that techniques which
simulate natural events can be employed to simultaneously
satisfy many of the interests of these various groups.
In keeping with the conference theme of water levels in
Great Lakes wetlands, I will omit from consideration here

those management techniques that are not related to water
levels.

I thank J. Bowlby and O. Schmitz for helping me
to better organize my thoughts for this paper. T. Nudds
reviewed an earlier draft.

MANAGEMENT FOR WHAT?

Many Great Lakes marshes have been lost to drainage
and in heavily settled areas this loss may range from
50 to 75 percent (Whillans, 1982). Many marshes that
remain do so only because of their economic return to
the landowners. Thus, one reason for management has been
to maintain or increase this economic return in order
to forestall the drainage of the marshes for agriculture
or other economic uses. If wetland preservation is an
objective, however, it may be risky to rely too heavily
on an economic basis for preserving wetlands because such
monetary approaches are uncertain and are seldom truly
long term. Moreover, they may seriously underestimate
the value of wetlands to future generations. The problems
in estimating the economic value of coastal wetlands are
discussed by Shabman and Batie (1979). Wetland preservation
should not have to rely on economics, for clearly these
systems are worthy of preservation for the sake of future
generations alone.

Water levels, nutrients and the frequency of natural
disturbance are possibly the main factors influencing
wetlands (Keddy, 1983). All of these factors have been
influenced by man to some degree. For example, in the
Great Lakes, humans have likely influenced both water
levels and the degree to which they fluctuate and have
also modified the frequency of natural disturbances like
wildfires. Muskrats can also influence vegetation, and
man has altered the frequency and amplitude of their population
changes as well. In addition, the introduction of exotic
plants and animals (e.g., purple loosestrife Lythrum salicaria,
carp Cyprinus carpio) have resulted in some marshes having
a far-from-natural flora and fauna (Weller, 1978b). Separating
natural and human-induced changes in wetlands can be difficult
because of indirect human effects (Keddy, 1983), but in
those wetlands where man has clearly altered the natural
community, management can be used to restore them.

A rather common objective of various interest groups
is the restoration, maintenance or improvement of habitat
quality, but habitat quality is inevitably defined by
the values and management objectives of these groups (Harris
et al., 1983). The quality of a habitat can be defined

for a single species by determining its habitat requirements, but determining the quality of a habitat for an entire community is more difficult. Reviewing the criteria that have been employed to quantitatively assess the value of different habitats, Margules and Usher (1981) concluded that diversity was the most used criterion (see also Asherin et al., 1979). Such an index of habitat quality seems well suited to a multi-species/community approach. It is clear that a single-species approach to improving, or even maintaining, habitat quality may have led to past conflict among groups interested in the production of different taxa.

A community approach to evaluating and managing habitat quality is not a panacea for the disagreements between interest groups although it is a step in reducing them. For example, if a group favors the production of a certain species and it appears that 10 percent more can be produced if a wetland is managed for this single species rather than for the marsh community, then a conflict might still exist. Fortunately, it appears that at least for vertebrates, management for habitat diversity may also be highly appropriate for most single species (and therefore little conflict may persist). Numerous workers have observed the positive correlation between waterfowl abundance, the number of waterfowl species and habitat heterogeneity or interspersion (e.g., Leopold, 1933; Griffeth, 1948; Steel et al., 1956; Steenis et al., 1959; Weller and Spatcher, 1965; Weller and Fredrickson, 1974; Weller, 1975). The findings of these studies have recently been experimentally confirmed by Kaminski and Prince (1981) and Murkin et al. (1982). Aside from ducks, a positive relationship between habitat diversity and bird abundance or species richness occurs in a variety of habitats (e.g., MacArthur, 1964; Pearson, 1977; Franzreb and Ohmart, 1978; Rice et al., 1983 and references therein). Harris et al. (1983) found that the overall avian community in a Lake Michigan wetland also exhibited this same trend. Common gallinules (Gallinula chloropus) were most abundant in Lake Erie marshes with a high degree of habitat diversity/interspersion (Brackney and Bookhout, 1982). Muskrat populations seem to reach their highest levels in the heterogeneous habitat that occurs when emergent vegetation and open water occupy equal areas within a marsh (Weller and Spatcher, 1965; Weller and Fredrickson, 1974). Because the fish community of marshes is composed of a diversity of species (Scott and Crossman, 1975), it seems likely that a community approach to marsh management will have similar benefits for fish. Weller (1978b) and Weller and Fredrickson (1974) suggested that high species richness can often be equated with high productivity within a marsh or a complex of marshes. Management then, need not always be for species A at the expense of species B. I believe that a single-species approach to management emphasizes such conflicts while a multi-species/community approach emphasizes commonalities.

MARSH DYNAMICS

A consideration of vegetational response to water level fluctuations is, of course, beyond the scope of this paper. However, I wish to briefly emphasize the dynamic and often cyclical nature of marshes as it pertains to management for wildlife. Many marshes exhibit cyclical vegetation changes in response to changes in water levels (Harris and Marshall, 1963; Weller and Spatcher, 1965; Weller and Fredrickson, 1974; Stuckey, 1975; Van der Valk and Davis, 1978, 1980; Van der Valk, 1981; Keddy, 1983) and many wetlands probably are maintained in a diverse and productive state because of some such change or other periodic disturbance (Kadlec, 1962; Harris and Marshall, 1963; Van der Valk and Davis, 1978; Weller, 1978a,b). Stable conditions, on the other hand, can be detrimental to some wetland vegetation (Uhler, 1944; Martin, 1953; Weller, 1975, 1978b).

Consider this cycle of vegetation change at the stage when an open (i.e., devegetated) marsh has recently experienced a major reduction in water level (a "drawdown") and mudflats are exposed where formerly there was open water. Many plant species require mudflats or very shallow water for their seeds to germinate (Martin, 1953; Weller, 1975, 1978b), and this germination often results in a densely vegetated marsh. The duration of this dense vegetation stage is reduced by deep flooding, intense muskrat grazing (Errington et al., 1963; Weller and Spatcher, 1965; Weller and Fredrickson, 1974) and sometimes wildfires (Ward, 1968). Eventually, an open (lakelike) stage may result and persist in this state until the next drawdown. Other studies have also considered the effects of drawdowns or flooding on marshes (Kadlec, 1962; Harris and Marshall, 1963; Meeks, 1969; Van der Valk, 1981; Pederson, 1981; Farney and Bookhout, 1982; Murkin, 1983). The response of aquatic invertebrates to changes in water levels or vegetation has been assessed by Kadlec (1962), Whitman (1974), Voigts (1976), Nelson (1982) and Murkin (1983) among others, but more research is required into the relationships of invertebrates to marsh dynamics (Weller, 1978b). Different animal and plant species may peak in abundance at various stages in this cycle, but frequently animal abundance and diversity are highest during the intermediate phase of this cycle when habitat heterogeneity is high (e.g.,

Beecher, 1942; Weller and Spatcher, 1965; Weller and Fredrickson, 1974). Considering the plant community per se, both Stuckey (1975) and Van der Valk and Davis (1980) concluded that fluctuating water levels increased the diversity of marsh plant species.

HOW TO MANAGE?

The flora and fauna of wetlands have been subjected to these natural fluctuations and perturbations (i.e., water levels, muskrats and fire) over evolutionary time. Therefore, it seems likely that management procedures that closely mimic these natural events will result in typical marsh communities. Weller (1978b, p. 275) reviewed the procedures employed to manage marshes and concluded that manipulating natural processes is not only more ecologically and economically sound than artificial techniques, but that the former approach is likely to have benefits for most plants and wildlife while the latter approach is more species-specific.

Clearly, one way to naturally manage the marsh community is to manage water levels. Although some water level control is currently exercised on some of the Great Lakes (Francis et al., 1979), we are, of course, considering dyked marshes here. Manipulation of water level in a dyked marsh can be used to offset these alterations of lake levels or other human effects. In addition, dykes can be used to create wetlands, or to preserve marshes in years when fluctuations in lake levels would eliminate them. Formerly, marshes shifted laterally about some mean shoreline in response to changes in lake levels. These irregular long-term changes in Great Lakes levels can be 1.5 m (Francis et al., 1979), so great lateral shifts in wetland location would occur. However, in some areas today where low-lying agricultural land abuts lakes, dykes have been built to maintain this land for agriculture even in years of high water, so no inland shift of marshes can occur. The marsh outside these dykes is flooded out at high water, and lake-like conditions prevail there until the water drops and vegetation can re-establish in shallow water or on mudflats. In such areas, we may effectively lose all undyked wetlands in high water years.

Water level control has several other advantages. Not only can it be used to increase plant and animal diversity within a marsh, but if at least some neighbouring marshes are dyked, then different water levels can be maintained in each. This would produce a complex of marshes in different successional phases and thus provide an even wider variety of habitats for plants and animals than might occur naturally.

Although dyking does have the advantages considered above, it is certainly not without disadvantages. The most obvious drawback is the reduced availability of marshes for fish feeding, spawning and nursery habitat (Francis et al., 1979). This is a serious problem and if a true community approach to marsh management is to be achieved, the importance of marshes to fish, and of fish to marshes, cannot be ignored. One possibility lies in the use of water control structures that allow fish passage. A second possibility is to determine the pattern of fish movement and to open up water control structures at times when fish move in and out of the marsh (Lapointe, 1983). The complete integration of fish into the management of dyked wetlands is a challenging problem and one that I believe is particularly worthy of future exploration.

Another aspect of water level control that would benefit from additional study is the reduction in mixing of wetland and littoral waters. Dyking interrupts the export of detritus into the lake from marshes and reduces the opportunity for marshes to remove sediments and nutrients from lake water (Francis et al., 1979).

A community/multi-species oriented alternative that may still allow many of the benefits of water level control, but better integrate these concerns about fish and nutrients, would be to dyke a marsh but allow its water level to fluctuate with the lake. Then, only in those years when a drawdown or deep flooding was really needed to improve the marsh would the water control structures be closed and hydraulic connectivity lost. In these years, the marsh would not be available to fish (unless fishways were provided) but if these manipulations improved habitat quality, the fish community might benefit in the long run.

A second way to naturally manage the marsh community is by altering muskrat populations or their impact on vegetation. Muskrats can affect marshes by cutting emergent vegetation for food or to build lodges. If, in fall, a muskrat cuts off the stalks of a cattail (Typha spp.) plant, for example, and water levels subsequently rise to cover the cut ends, the rhizome is separated from its supply of atmospheric oxygen. The rhizome may contain enough oxygen to overwinter but in spring its oxygen needs increase (Sale and Wetzel, 1983). If a growing shoot does not reach the surface of the water soon enough the plant dies. Heavy muskrat cutting can also open up dense cattail stands even if water levels do not rise. Overwinter survival of muskrats is affected by water levels (Meeks,

1969). Therefore, water level manipulation can also be used to indirectly manage marshes by altering muskrat populations or by modifying the impact of their cutting on vegetation.

A third way to naturally manage the marsh community is to employ fire in conjunction with water level manipulation. Uhler (1944) and Hoffpauer (1968) described how marsh fires during severe drawdowns can sometimes directly kill vegetation by consuming rhizomes. Like cutting, fire can also merely remove the parts of the plant above water, and if the ends of these burnt stalks are submerged, the plant may die (Ball, 1984). Fire is a natural way of managing a marsh, but its effects on marshes are not well understood and further research is needed (Ward, 1968; Weller, 1978b; Kaminski and Prince, 1981).

THE NEED FOR EXPERIMENTS

If a large number of factors (like water levels, muskrat densities and nutrient availability, for example) all change simultaneously, any cause-and-effect relationships are obscured. Such is the case even if we are observing a dynamic system because these factors are often cross-correlated. Observational studies on marshes have provided a wealth of information, but as Macnab (1983) noted, it is only by experimentation that we can examine what these factors do. This is why a statement of causality is usually reserved for manipulative experiments (Tabachnick and Fidell, 1983). In single-species management, it is often not too difficult to evaluate its effects on that one species by observation. However, if we take a community approach to management, and especially if we employ natural techniques, many factors change simultaneously while many species respond to these changes or to the responses of other species. To most effectively employ natural management techniques, we must understand the direct and indirect consequences of our actions. We do not have to firmly establish all the cause-and-effect relationships before we can take a natural approach to the management of the marsh community, but such knowledge cannot help but improve management.

I have tried to cite experimental work when possible, but much of our knowledge is based on observational studies. I do not mean to suggest that all studies must be experimental to be correct, but merely, that to answer certain questions only an experiment will do. Salt (1983) commented that field experiments are highly valued despite their inherent difficulties in design and execution. These difficulties are often not overcome and as a result, many of the few experimental studies of wetlands that have been performed have little if any replication and therefore lack sensitivity (Eberhardt, 1970). Thus, one direction of future research should be toward designing replicated experiments to determine the relationships between wetland variables.

I will briefly summarize three studies that were experimental, replicated, and which examined the relationships between various components of the community (here, vegetation, invertebrates, habitat structure, and waterfowl) to get answers to questions that elude observational studies. Consider the correlation that managers have long observed between waterfowl abundance and habitat diversity (often simply the interspersion of emergent vegetation with open water). Numerous observational studies have documented this correlation (e.g., Leopold, 1933; Griffeth, 1948; Steel et al., 1956; Steenis et al., 1959) but only an experiment can determine causality if intercorrelations exist. Ducks may respond directly to this habitat structure or perhaps they respond to some correlated factor like the distribution of aquatic invertebrates. To test the hypothesis that both habitat structure and food are important, Kaminski and Prince (1981) performed an experiment in which these two factors were not correlated. They controlled water levels so that this was not a confounding factor. Kaminski and Prince created circular areas of open water (0.1 ha in size) in a stand of dense vegetation by one of two techniques (mowing or rototilling). They varied the number of circular openings to create areas which differed in the ratio of water to vegetation. Basin treatment differentially affected the aquatic invertebrate foods of ducks, so ducks were presented with a situation in which food was uncorrelated with habitat structure. Their findings indicated that habitat structure was more important in affecting the distribution of ducks than was food.

It is important that tests of hypotheses not only have replication, but be replicated. Murkin et al. (1982) determined the effects of habitat configuration and food on duck distribution using a different experimental design than Kaminski and Prince (1981). Murkin et al. (1982) manipulated the ratio of emergent vegetation on open water by cutting channels in a checkerboard pattern through a dense stand of emergent vegetation. The results of their study confirmed the findings of Kaminski and Prince (1981), specifically, that ducks did respond directly to habitat structure (as indexed by the ratio of cover to water). In addition, Murkin et al. (1982) found that duck distribution was affected by invertebrates.

Building on the studies of Kaminski and Prince (1981) and Murkin et al. (1982), I conducted an experiment to examine the influence of habitat structure and invertebrates on waterfowl, and in addition, the effects of two natural management techniques on vegetation and invertebrates. The ratio of cover to water was not considered as it was in these two previous studies but instead, looked at another aspect of habitat heterogeneity, the size of the patches of open water. Two techniques that simulated natural events (fire and cutting, the latter simulating muskrat activity) were used to create patches of open water (size 0.02, 0.09, and 0.15 ha) in dense cattail. Prior to these manipulations, the water level was drawn down. Cattail was cut and burned over the ice in the winter and then in spring, the water level was raised to cover the cut and burnt ends of the cattail stalks. Few studies on the effects of cutting or burning have been experimental, and it is this lack of experimental control that may explain their lack of agreement on the effects of these treatments. Water levels were chosen and maintained within 2 cm in order to eliminate the confounding effects of within-season water level variations.

Cutting resulted in significantly greater cattail mortality than burning although both did kill cattail if the stalks were submerged in spring. The contrast between the response of aquatic invertebrates to these management techniques in the first and second years underscores the need, not only for experiments, but for longer term studies. Because of the dynamic nature of marshes (both within and over years), long term experiments may reveal trends that are not distinguishable in short-term studies. Short-term studies may be difficult to interpret correctly because they represent "snapshots" of a changing system. Although my study was not long term, I did observe significant yearly differences in the aquatic invertebrate response to the experimental changes in water levels and to the cutting and burning treatments. In the first year of flooding, burned areas had approximately double the invertebrate biomass of open water areas, while mowed areas had between three and four times the biomass of open water sites. However, in the second year of high water after drawdown, there were no significant differences among these three areas. The detritus from emergent plants is directly or indirectly the food source for most of these invertebrates and, as Nelson (1982) hypothesized, it appears that managers can increase invertebrate abundance by altering emergent detrital inputs to marshes. These invertebrates affected mallard (Anas platyrhynchos) distribution as did habitat structure (here, opening size).

There are, of course, other experimental studies on various aspects of wetland ecology. The three studies that I have taken as examples, however, illustrate the need for long-term experimental studies, with adequate replication, that evaluate the relationships between the components of the system.

CONCLUSION

I believe that a community/multi-species approach to marsh management will both minimize conflicts between interest groups and maximize benefits for most plants and animals. Natural techniques are particularly appropriate for this community approach to management. Alone or in conjunction with fire or muskrat activity (real or simulated by mowing), water level manipulation is an ideal way to manage marshes. Dykes can be used to create or restore marshes, but a true community approach to management is required to maximize the benefits to all species.

In using natural techniques of management, many results we observe are indirect. Long-term, experimental, interdisciplinary studies are needed both to understand the effects of such management, and to understand the natural dynamics that we seek to duplicate.

DISCUSSION

Prepared By

Jeffrey W. Nelson
Ducks Unlimited Canada
Winnipeg, Manitoba R3T 2E2
and
Ted Gadawski
Ducks Unlimited Canada
Barrie, Ontario L4N 4Y8

A common theme running through many of the papers presented at this colloquium has been to develop an understanding of how coastal wetlands function. Perhaps most importantly, a better understanding of wetland function will enhance our ability to predict impacts of various cultural practices. Whether water levels fluctuate by design or through natural variation, the same forces function to shape wetland communities. If water levels can be controlled, the effectiveness of wetland management programs is directly proportional to how predictably the wetland community will change. Thus, ultimately, wetland managers stand to gain significantly from a better understanding of how wetland communities function under varying water levels.

All three "community approaches" to wetland management suggested by Ball (1984) reduce in simplest form to wetland vegetation management. Obviously, water levels must be carefully controlled to ensure that any one of the techniques is effective. Although we all recognize that diking of coastal wetlands interrupts interactions between adjacent lacustrine and palustrine systems, management as proposed by Ball (1984) requires wetland isolation for water level control. Our challenge is to develop diking systems and control structures that allow fish passage and flushing action between managed coastal wetlands and adjacent lakes. In that way, a productive and diverse wetland community can be maintained during periods of both high and low lake levels. Water level changes in managed wetlands are controlled, as opposed to stabilized, if diked areas can be established.

The two studies (Kaminski and Prince, 1981; Murkin et al, 1982) cited by Ball (1984), in addition to his own study, all demonstrate bird response to habitat structure. An important question, which no study has addressed, is why do wetland bird communities respond to water/vegetation structure as documented. We need to develop a conceptual framework of how opening size, emergent vegetation, and water levels interact and provide proximate cues to wetland avifauna selecting habitat. Nelson and Kadlec (in press) have proposed a conceptual framework attempting to link prairie wetland ecology and waterfowl breeding biology that could serve as a basis for understanding wetland function more generally in coastal Great Lakes wetlands.

Finally, I believe better communication is necessary both among wetland research biologists and between researchers and those attempting to apply results to practical decision making. Although many here have emphasized the bewildering complexity of wetland systems and have stressed the need for continued long-term research, the sad fact is that important decisions about the wetland base are, by necessity, being made today. That fact underlines the need for all of us to ensure that research results are efficiently and effectively transmitted. There is a real need for technical information to be condensed and presented in an easily read and understood format for wetland specialists other than research biologists. By way of example, a prime objective of the Marsh Ecology Research Project, being conducted at the Delta Waterfowl and Wetlands Research Station in Manitoba, is to ensure that an increased understanding of marsh function developed by that research program is effectively transferred to wetland managers. Wetland colloquiums, like the one we're attending here, foster better communication among those interested in understanding, developing, and protecting our remaining wetland base.

LITERATURE CITED

Asherin, D.A., H.L. Short, and J.E. Roelle. 1979. Regional evaluation of wildlife habitat quality using rapid assessment methodologies. Trans. N. Am. Wildl. and Nat. Resour. Conf. 44:424.

Ball, J.P. 1984. Habitat selection and optimal foraging by mallards: A field experiment. MSc. Thesis, University of Guelph, Guelph, Ontario.

Beecher, W.J. 1942. Nesting birds and the vegetative substrate. Chicago Ornithological Soc., Chicago, IL. 60 pp.

Brackney, A.W. and T.A. Bookhout. 1982. Population ecology of common gallinules in southwestern Lake Erie marshes. Ohio J. Sci. 82(5):229-237.

Eberhardt, L.L. 1970. Correlation, regression, and density-dependence. Ecology. 51:306-310.

Errington, P.L., R. Siglin, and R. Clark. 1963. The decline of a muskrat population. J. Wildl. Manage. 27:18.

Farney, R.A. and T.A. Bookhout. 1982. Vegetation changes in a Lake Erie marsh (Winous Point, Ottawa County, Ohio) during high water years. Ohio J. Sci. 82(3):103-107.

Francis, G.R., J.J. Magnuson, H.A. Regier, and D.R. Talhelm (eds.). 1979. Rehabilitating Great Lakes Ecosystems. Tech. Rep. No. 37 Great Lakes Fishery Commission, Ann Arbor, MI.

Franzreb, K.E. and R.D. Ohmart. 1978. The effect of timber harvesting on breeding birds in a mixed coniferous forest. Condor. 80:431-441.

Griffeth, R.E. 1948. Improving waterfowl habitat. N. Am. Wildl. Conf. Trans. 13:609-617.

Harris, H.J., M.S. Milligan, and G.A. Fewless. 1983. Diversity: Quantification and ecological evaluation in freshwater marshes. Biol. Conservation. 27:99-110.

Harris, S.W. and W.H. Marshall. 1963. Ecology of water-level manipulations on a northern marsh. Ecol. 44(2):331-343.

Hoffpauer, C.M. 1968. Burning for coastal wetland habitat. In: J.D. Newsom (ed.), Proc. Symp. Marsh and Estuary Manage. pp. 134-139.

Kadlec, J.A. 1962. Effects of a drawdown on a waterfowl impoundment. Ecol. 43(2):267-281.

Kaminski, R.M. and H.H. Prince. 1981. Dabbling duck and aquatic macroinvertebrate responses to manipulated wetland habitat. J. Wildl. Manage. 44:1-15.

Keddy, P.A. 1983. Freshwater wetlands human-induced changes: Indirect effects must also be considered. Env. Manage. 7(4):299-302.

Lapointe, G. 1983. Fish movements and predation on invertebrates in the Delta marsh. p. 15 in: Delta Waterfowl Research Station 1983 Annual Report. 32 pp.

Leopold, A. 1933. Game Management. Chas. Scribner and Sons, New York, NY. 481 pp.

MacArthur, R.H. 1964. Environmental factors affecting bird species diversity. Am. Nat. 98:387-396.

Macnab, J. 1983. Wildlife management as scientific experimentation. Wildl. Soc. Bull. 11:397-401.

Martin, A.C. 1953. Improving duck marshes by weed control. U.S. Fish Wildl. Serv. Circ. 19. 49 pp.

Margules, C. and M.B. Usher. 1981. Criteria used in assessing wildlife conservation potential: A review. Biol. Cons. 21:79-109.

Meeks, R.L. 1969. The effect of drawdown date on wetland plant succession. J. Wildl. Manage. 33(4):817-821.

Murkin, H.R. 1983. Responses by aquatic macroinvertebrates to prolonged flooding of marsh habitat. Ph.D. Thesis, Utah State University, Logan, UT.

Murkin, H.R., R.M. Kaminski, and R.D. Titman. 1982. Responses by dabbling ducks and aquatic invertebrates to an experimentally manipulated cattail marsh. Can. J. Zool. 60:2324-2332.

Nelson, J.W. 1982. Effects of varying detrital nutrient concentrations on macroinvertebrate abundance and biomass. MSc. Thesis, Utah State University, Logan, UT.

Pearson, D.L. 1977. A pantropical comparison of bird community structure on six lowland forest sites. Condor. 79:232-244.

Pederson, R.L. 1981. Seed bank characteristics of the Delta marsh, Manitoba: Applications for wetland management. In: Richardson, B. (ed.), Selected Proceedings of the Midwest Conference on Wetland Values and Management. June 17-19, 1981, St. Paul, MN. 660 pp.

Rice, J.R., R.D. Ohmart, and R.W. Anderson. 1983. Habitat selection attributes of an avian community: A discriminant analysis investigation. Ecol. Monog. 53(3):263-290.

Sale, P.J. and R.G. Wetzel. 1983. Growth and metabolism of Typha species in relation to cutting treatments. Aquatic Botany. 15:321-324.

Salt, G.W. 1983. Roles: Their limits and responsibilities in ecological and evolutionary research. Am. Nat. 122:697-705.

Scott, W.R. and E.J. Crossman. 1975. Freshwater fishes of Canada. Bull. 184 Fisheries Research Board of Canada. 966 pp.

Shabman, L.A. and S.S. Batie. 1979. Estimating the economic value of coastal wetlands: conceptual issues and research needs. Sea Grant Project Paper VPI-SG-79-08. Virginia Tech., Blacksburgh, VA.

Steel, P.E., P.D. Dalke, and E.G. Bizeau. 1956. Duck production at Gray's Lake, Idaho 1949-1951. J. Wildl. Manage. 20:279-285.

Steenis, J.H., L.P. Smith, and H.P. Cofer. 1959. Studies on cattail management in the northeast. Trans. Northeast Wildl. Conf. 149-155.

Stuckey, R.L. 1975. A floristic analysis of the vascular plants of a marsh at Perry's Victory Monument, Lake Erie. The Michigan Botanist. 14:144-166.

Tabachnick, B.G. and L.S. Fidell. 1983. Using Multivariate Statistics. Harper and Row, New York. 509 p.

Uhler, F.M. 1944. Control of undesirable plants in waterfowl habitats. N. Am. Wildl. Conf. Trans. 9:295-303.

Van der Valk, A.G. 1981. Succession in wetlands: A Gleasonian approach. Ecol. 62:688-696.

Van der Valk, A.G. and C.B. Davis. 1978. The role of seed banks in the vegetational dynamics of prairie marshes. Ecology. 59:322-335.

Van der Valk, A.G and C.B. Davis. 1980. The impact of a natural drawdown on the growth of four emergent species in a prairie glacial marsh. Aquatic Botany. 9:301-322.

Voigts, D.K. 1976. Aquatic invertebrate abundance in relation to changing marsh vegetation. Am. Midl. Nat. 95:313-322.

Ward, P. 1968. Fire in relation to waterfowl habitat of the Delta marshes. Proc. Tall Timbers Fire Ecol. Conf. 8:254-267.

Weller, M.W. 1975. Studies of cattail in relation to management for marsh wildlife. Iowa State J. Res. 49(4):383-412.

Weller, M.W. 1978a. Wetland habitats. In: Wetland Functions and Values: The State of Our Understnading. P.E. Greeson, J.R. Clark and J.E. Clark (eds.)., Proc. Natl. Symp. on Wetlands.

Weller, M.W. 1978b. Management of freshwater marshes for wildlife. In: Freshwater Marshes: Ecological Processes and Management Potential. R.E. Good, D.F. Whigham, and R.L. Simpson (eds.)., Academic Press, New York, NY.

Weller, M.W. and L.H. Fredrickson. 1974. Avian ecology of a managed glacial marsh. The Living Bird. 12:269-291.

Weller, M.W. and C.S. Spatcher. 1965. Role of habitat in the distribution and abundance of marsh birds. Iowa Agric. and Home Econ. Exp. Stn. Spec. Rep. 43. 31 pp.

Whillans, T.H. 1982. Changes in marsh area along the Canadian shore of Lake Ontario. J. Great Lakes Res. 8(3):570-577.

Whitman, W.R. 1974. The response of macro-invertebrates to experimental marsh management. Ph.D. Thesis, University of Maine, Orono, ME.

INDEX